D0365003

IS MY CELL PHONE BUGGED?

IS MY CELL PHONE BUGGED?

EVERYTHING YOU NEED TO KNOW TO KEEP YOUR MOBILE CONVERSATIONS PRIVATE

KEVIN D. MURRAY

Published by Emerald Book Company
Austin, TX
www.emeraldbookcompany.com

Distributed by Emerald Book Company

For ordering information or special discounts for bulk purchases, please contact Emerald Book Company at PO Box 91869, Austin, TX 78709, 512.891.6100.

Design and composition by Greenleaf Book Group LLC and Bumpy Design
Cover design by Greenleaf Book Group LLC

Publisher's Cataloging-In-Publication Data
(Prepared by The Donohue Group, Inc.)
Murray, Kevin D, 1950-
Is my cell phone bugged? : everything you need to know to keep your mobile conversations private / Kevin D. Murray. — 1st ed.
p. ; cm.
ISBN: 978-1-934572-88-7
1. Cell phones—Security measures—Popular works. 2. Personal communication service systems—Security measures—Popular works. 3. Eavesdropping—Popular works. I. Title.
TK5103.485 .M87 2011
384.5/3 2011923307

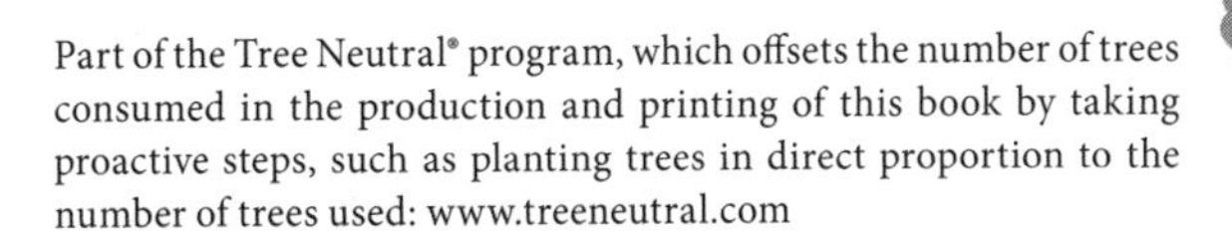

Part of the Tree Neutral® program, which offsets the number of trees consumed in the production and printing of this book by taking proactive steps, such as planting trees in direct proportion to the number of trees used: www.treeneutral.com

Printed in the United States of America on acid-free paper

11 12 13 14 15 16 10 9 8 7 6 5 4 3 2 1

First Edition

DISCLAIMER

To Charles Greenwald, Patricia Greenwald,
and Eugene Murray, who gave me the love, education,
and inspiration that made my career possible,
and to my family, Wendy, Rachel, and Garret, for their
support and encouragement while writing this book.

CONTENTS

FOREWORD

During the Cold War, eavesdropping on telephone communications was widespread and useful. Tapping the telephone lines leading into an opposition embassy usually would not reveal secret information, but it would often reveal with whom the embassy was in contact in the local community. That could result in more taps, which might then reveal hostile activity. An espionage practitioner often would have to leave his embassy and go to a public pay telephone to call his agent, because he suspected or assumed that his embassy lines were tapped.

Times have changed. For the average person, wiretapping is no longer a remote, government concern. Nearly everyone carries instant access to phone service, and tapping has become a *personal* concern. In fact, the term "wiretapping" is outdated. The new electronic surveillance is wireless. It is made possible by inexpensive software and inexpensive hardware, and it is supported by an extensive global communications infrastructure.

The focus of espionage has changed, too. Eavesdropping on a phone call is no longer the only type of attack. Text messages, call logs, and e-mails are also eavesdropping targets. The phone itself can be silently activated, turning it into a personal bugging device. These conversations, messages, and bits of data are automatically forwarded to the spy, who can be located anywhere Internet or cellular phone service is available.

Almost every person with a cell phone is a potential victim.

Even the spies have changed. Your spy is no longer hiding in a government embassy somewhere. He or she may be a person you know, a spouse, a significant other, an employer, a "friend," a colleague, a neighbor, or a landlord. It may be someone you don't know—a fraudster, an identity thief, a stalker, or any number of assorted criminals or perverts. All of these types of people have one thing in common: They can commandeer the communications tool you carry with you.

The following material deals with mobile communications security and technology in considerable detail. It is a valuable contribution to the security industry—and to everyone who wants to protect their personal privacy.

—Glenn Whidden, CIA–Clandestine Service (Retired); Technical Services Agency, Inc., Fort Washington, Maryland (USA); founder, Espionage Research Institute, Washington, DC

PREFACE

"You're just being paranoid."

It's a phrase that intimidates, shames, and scares. Too often, it sentences real victims of electronic surveillance to silent suffering. It's also a phrase that can reveal unflattering things about the speaker, who may simply be ignorant, shallow, or mean, and who sometimes shows a strong tendency to avoid reality. The fact is, other people cannot make your problems go away by telling you that they do not exist—and neither can you.

Life has taught all of us some valuable lessons: An ounce of prevention really is worth a pound of cure. Trust your instincts. And that noise you heard coming from your car's engine yesterday will not go away tomorrow; it *will* get worse.

Granted, some people really do have paranoia problems. But these people usually do not confess to having a specific fear about specific events, such as receiving odd text messages on their cell phone or hearing background noises that were never there before. They express their concerns in more general terms, such as "They know everything about me" or "It's been going on for years." Regardless, these people need kindness and medical help, not name calling.

The vast majority of us, however, lead busy lives. Thoughts of electronic surveillance and spying do not normally occur to us. We have too many *real* thoughts vying for our attention. There is neither time nor reason to dream up stories of omnipotent adversaries who know our every spoken word, thought, and move.

If thoughts of eavesdropping are new to you and you have a suspect with a motive in mind, *pay attention*. Your intuition is telling you that *something is wrong*. Too many "coincidences" have tipped your inner warning scale. Your subconscious alert is sounding a real alarm, just as surely as the smell of smoke reminds you of the food left burning on the stove.

Trust your judgment. Something *is* wrong.

My colleague, Gavin de Becker, a well-known authority on personal protection, explains this phenomenon very well in his book *The Gift of Fear*. He emphasizes the point about trusting your intuition:

> Park Dietz, the nation's leading forensic psychiatrist and an expert on violence, has noted that the case histories are "littered with reports, letters, memoranda, and recollections that show people felt uncomfortable, threatened, intimidated, violated, and unsafe because of the very person who later committed atrocious acts of violence."

Although the main topic of the book you hold in your hands is spyware on cell phones, you must also accept the fact that information leaks are not always caused by this type of technical surveillance. In fact, they are not always caused by any one form of espionage. You may be dealing with a situation in which your adversary has gleaned his/her information about you in several ways. To cover all such possibilities and *really* solve your concerns, you need to take the following steps:

- Examine *all* the ways your private conversations become public.
- Conduct tests to find the leaks.
- Follow recommendations that will keep this surveillance from happening again.
- Determine whether someone is harassing you into *thinking* that electronic eavesdropping is how he/she is getting information about you.

My goal is not simply to help you determine whether your cell phone is bugged. My ultimate goal is to help you solve your concerns about the loss of your privacy, no matter how it may have occurred.

What you will learn by following the directions in this book may surprise you. Questions that seem to have *no possible answer* often give way to clear answers when a *compound explanation* is considered.

Here is a simple example: A colleague at work says, "I see you went to the mall last night." How did she know? You told only one close friend during a phone call that you were going. Did your colleague follow you? Did she eavesdrop on your phone call?

The answer is actually much less nefarious: Your work colleague saw you break your eyeglasses yesterday. Since you are, in fact, wearing new glasses today, it was probably just a logical assumption on her part—an innocent conversation starter.

Things change, however, when the other person *knows your every move, every day.* Here is how you can tell the difference:

Murray's Rule of Thumb

- One incident is coincidence.
- Two incidences are suspicious.
- Three incidences are a confirmation.

In the end, you may find that your spy actually *did* know what you were thinking and doing—via a *compound explanation.* Government intelligence agencies uncover secrets this way all the time. Predicting future actions by cleverly factoring *several* bits of information together is a standard operating procedure.

The executives with whom I consult are often surprised when they learn that their spy *didn't* read their minds. Here is what really happens: A sharp spy collects the same bits of information that are available to the executive, often before this information reaches the executive for evaluation. The spy evaluates it as the executive would, only sooner, and comes to the same conclusions as the executive would, only faster. Hence the executive's lament, "This person knew what I was going to do before I decided to do it!" This phenomenon is more common than most people think.

Privacy invasion via wireless technologies is particularly scary. "How did they do it?" the victim asks himself, a response that is equal parts awe and fear. Electronic eavesdropping can seem like black magic, but the truth is, it's *not* magic—and a little education is all you need to put you back in control again.

You were smart to bypass the guesswork of the gadflies who pepper the Internet with *free* advice, and the alluring links to the detection and protection gadgets they secretly promote. Instead, you did some research and bought this book. Thank you for your trust. I am going to do my very best to help you address your concerns and solve your problem.

Is My Cell Phone Bugged? is one book in my series of *personal counterespionage* works aimed at helping people regain their privacy and the security of their valuable information. The series covers specific problems, one topic at a time.

What you are about to read is very similar to an actual private consultation with a certified, professional security consultant. The advice presented here evolved from my experience in solving my clients' real-life problems over a period of three decades. These privacy solutions have been stress-tested—frequently. We know they work. So we're going to apply them to your particular situation.

Let's get started by addressing *your* concerns and solving *your* problems.

ACKNOWLEDGMENTS

The book you see now was refined, enhanced, and polished with the assistance of:

- Aaron Bowler, online-stopwatch.com, who custom designed our special on-line transmissions timer
- Joseph E. Canone, Jr., director, Pinnacle Investigations
- Kevin A. Cassidy, security director, Thomson Reuters
- Mark J. Cheviron, security director, ADM
- Nicole Cowley, a brilliant writer who should write her own book
- Rachel A. Girardin, with her talents for design and simplification
- Garret L. Girardin, with his talent for honest appraisal
- Heather G. Greig, who always finds my grammar mistakes no matter how hard I try to conceal them

- Edward Lee, CPP, retired U.S. diplomat, federal agent, and author of *Staying Safe Abroad*
- William Milley, whose creative ways of explaining techie stuff make it fun and interesting
- Gordon L. Mitchell, Ph.D., CPP, Future Focus/eSleuth—TSCM and Computer Forensics
- Annemarie Mudd, ace proofreader
- Wendy H. Murray, J.D., F.C.I., my wife
- Philip Jan Rothstein, FBCI, Rothstein Associates, Inc., DisasterRecoveryBooks.com
- Mr. Alexander C. Sparaco, CPP, director, Baker Street Investigative Services
- "Pete" Landon L. Smith (deceased), inventor, mentor, and good pal
- Ruth Smith, ace proofreader
- Andrew M. Tripp, a brilliant TSCM tech and voice of reason
- Glenn H. Whidden, CIA—Clandestine Service (retired), Technical Services Agency, Inc.

Their advice and skills have helped me make cell phone information privacy readily available and easily understandable. My sincere thanks to all of you.

INTRODUCTION

The level of privacy you experience is *inversely proportional* to the amount of **time** and **money** an eavesdropper can devote to spying on you, plus the **value** of what he/she can learn through surveillance. This **value** may be anything from multimillion-dollar business secrets to a divorce strategy to the desire to satisfy prurient interests.

Unfortunately, it is not only your cell phone that puts your privacy at risk. Consider the many other ways you communicate without wires. If you are reading this book, you probably use at least two of these wireless devices regularly:

- www))) Cellular phone (GSM, CDMA, with SMS/MMS capabilities)[1]
- www))) Home cordless phone (analog, digital spread-spectrum, DECT, etc.)

1 Note that for simplicity, all wireless public telephone system phones are referred to as "cellular/cell phones." Short-range, wireless extensions to business and residential telephone lines are referred to as "cordless phones."

- www)) In-transit airline and ocean liner phone (Picocell via satellite link)
- www)) Personal microcell phone (used at some corporate campuses)
- www)) Satellite phone
- www)) VoIP Wi-Fi phone software (on laptops, cell phones, or tablet devices)
- www)) Bluetooth® cordless phone, speakerphone, and accessories
- www)) Cordless headset[2]

Each of these technologies has its own set of vulnerabilities. Fortunately, the privacy precautions and security techniques provided here can be applied to counter just about all of them.

To enhance your cell phone privacy, we need to take the following steps:

- **Demystify the technical aspects of wireless communications.** You need to understand what can happen, what can't happen, and what *is* happening.
- **Separate the facts from the fiction.** For example: No, there is no phone number you can dial to see if your phone is bugged or wiretapped. This is an urban legend that at one point had a tiny grain of truth to it—before the phone companies changed over to electronic switching in the 1970s and 1980s.
- **Consider some of the *not-so-obvious* possibilities.** While a bugged cell phone is a very real possibility,

2 The www)) you see above and in the following chapters indicates that definitions and additional information are available about technical topics, terms, and acronyms. Visit the *Is My Cell Phone Bugged?* companion Web page (http://www.spybusters.com/Cell911.html) to access this extra information.

it is not the only way someone can learn about your activities.

- **Mix in some common sense.** Surprise! Searching for an eavesdropping device is *not* the best first step toward solving the problem.
- **Create security checklists.** Following these suggestions will reduce your security loopholes and ensure increased privacy in your communications.

The good news is, all this can be accomplished in a very cost-effective and nontechnical manner.

First, let's ask a few important questions.

HOW LIKELY IS IT THAT SOMEONE IS LISTENING TO YOUR PHONE CALLS?

It's hard to say. The type of phone you use and the importance of your calls (as perceived by nosy neighbors, spouses or significant others, business competitors, law enforcement agencies, etc.) contribute to the likelihood of your call being a target for eavesdropping. Also, eavesdropping is time-consuming and it is costly. These elements all factor into the probability of you being a target.

Also keep in mind that eavesdroppers thrive on information. To them, information means either power, money, or some sort of enjoyment. That's their reward. Thus, our obvious first tip is . . .

TIP ▪ The less information you give them, the sooner they will get bored and move on.

WHAT ABOUT LAW ENFORCEMENT?

For law enforcement agencies to legally eavesdrop, they must follow certain procedures laid out by law. Often they must also follow additional strict internal guidelines before they are allowed to tap into a phone call. Unless you are *really* interesting from their perspective, you will not be on their list of usual suspects—and so they should not be on *your* list of usual suspects. The days of rogue phone taps by cops and "my buddy at the phone company" are pretty much over.

By the way . . . if you *are* concerned that you're the target of a legal, court-ordered wiretap, assume that your communications may be intercepted. The Communications Assistance for Law Enforcement Act of 1994 (CALEA) mandated that all U.S. phone companies modify their equipment to make legal wiretapping easy. The end result is that, with a few keystrokes, any call may be easily monitored. This monitoring *does not* change the physical or electrical characteristics of the phone (wired or wireless); thus it is undetectable by the user no matter what type of mobile or landline communications device he/she is using.

If surveillance by a government agency is your concern, put this book down and slowly back away from the cash register. The information given here cannot help you. You need to read Dr. Dorothy E. Denning's *Wiretap Laws and Procedures: What Happens When the U.S. Government Taps* ((www *a Line*. Although it was written some years ago, much of the information in this white paper remains pertinent today.

Come to think of it, you should also read the "Legal Issues" chapter of this book and learn about the current laws referenced there. OK, you can pick up this book again and proceed to the cash register.

"Illegal eavesdropping is outrageous. There ought to be a law!"

www)) There *are* laws—several of them. Eavesdropping by private individuals on any telephone call is illegal in the United States, with very few exceptions. But do these laws help? A little. Still, the deck is stacked against you. Analyzing the key issues involved in protecting your privacy will help you understand why:

- **Availability.** Bugging devices and spyware that would have made James Bond and Q giddy just ten years ago are now commonplace items. Today, Bond's toys are available to everyone. If a would-be spy has a computer, a credit card, and Internet access, he/she can shop from home.
- **Low cost of entry.** Bond and Q had unlimited government budgets, which was crucial since spy gadgets used to be very expensive. Today, however, GSM bugs, which can remotely monitor from anywhere there is phone service, are sold on eBay for less than $25. Cell phone spyware is also moderately priced.
- **Low, low probability of detection.** Even if a bug or spyware is discovered, there is usually no evidence

linking the criminal to the crime. From the eavesdropper's viewpoint, the chances of being caught are close to nil. Thus, laws provide little deterrence and may even seem irrelevant.

- **Low, low, low probability of prosecution.** On the off chance that an eavesdropper *is* caught, the probability of him/her being prosecuted is pretty slim. Law enforcement and the courts are overwhelmed with bigger fish to fry. Given the workload of more serious crimes that the legal system has to handle, personal eavesdropping doesn't carry much weight. The proof of this can be seen on Internet auction sites and spy shops, where sellers of illegal bugging devices enjoy an open and flourishing trade.
- **Low, low, low, low probability of meaningful punishment.** Let us assume, for the sake of argument, that an eavesdropper *is* caught and *is* successfully prosecuted. Still, eavesdropping is not viewed as a particularly heinous crime (one exception: using spycams to peep on minors in the bathroom). A fine and possibly a probation sentence are all that can be expected for punishment. A trip to the Gray Bar Motel is unlikely. In one recent case, a former Wall Street broker was sentenced to no jail time and only a $500 Fine, even though he pleaded guilty to conspiracy to commit securities fraud by allowing others to eavesdrop on the firm's confidential squawk box conversations.[3]

3 Chad Bray, "'Squawk Box' Case Witness Receives No Prison Time," *The Wall Street Journal*, September 9, 2010.

While actual criminal statistics do not yet exist to support or refute these points, we can tell from news reports over the years that eavesdroppers have little to fear from the judicial system. Many examples of this are documented in the long-running online newsletter *Kevin's Security Scrapbook* (http://spybusters.blogspot.com). The legal system kicks in only after a crime is committed, but given the key issues listed above, laws do not have much *deterrence* value.

TIP ▪ You cannot depend on the law to protect you from electronic eavesdropping.

So, if you cannot count on the law to punish your snoop, what *can* you do? The answer is clear: *You* must take steps to protect yourself.

Naturally, you must become acquainted with the technology used for electronic surveillance. But even more important, the first step is to make a change in your mindset. *How*, *where*, and *when* you use your wireless communications devices is as important as choosing the right technology. Plus, you need to believe that eavesdropping *really* happens and that *real* people are doing it. Most people have a difficult time with this, even when faced with the evidence.

Because we cannot *see* eavesdroppers, there is a natural tendency to think there are no eavesdroppers. The result is that we use our communications tools as if they offer 100 percent privacy. Strange! One would not apply the same faulty logic to driving a car: We cannot see our next accident coming, but this does not mean we dispense with the seat belts.

Communications privacy is no less precarious. Precautions are necessary. In the following pages you will learn how mobile communications eavesdropping is conducted. Safety measures and security tips will be provided. Think of them as seat belts for your thoughts. Use them, and your communications will be a whole lot safer from eavesdropping.

THE TECHNOLOGY

The many portable communications devices we use have one thing in common: a radio signal. Your thoughts may start out as text or sounds, but ultimately, they are transported to your recipient as a radio signal traveling through space. Intercepting these signals is actually quite easy. Converting them back into sound, text, or video is the challenging part.

Varying amounts of instrumentation and expertise are required for successful reconstruction of the intelligence hidden within the signals. Some signals can be easily intercepted using commonly available radio receivers. Other signals require more specialized receiving equipment *and* complicated signal processing before the eavesdropping can begin. In the end, however, no signal is invulnerable to decoding. Success is just a matter of time, resources, and determination.

There are two secrets to maintaining communications privacy: Either make the decoding process *take so long* that it becomes not worth the time and effort to eavesdrop, or make the decoding process *so difficult* that it is beyond the capabilities of your particular eavesdropper. All the worthwhile advice and tips about the type of communications devices you should use must focus on these two goals. And using the right mode of transmission can go a long way toward instantly increasing communications privacy.

ANALOG VS. DIGITAL

Radio signals come in two basic types: analog and digital. ((www

Analog signals are the easiest targets for eavesdroppers. Wireless baby monitors are the most familiar example of this. Over the past couple of decades, neighbors have been inadvertently overhearing neighbors. Burglars listen for baby monitors when casing neighborhoods for empty homes. And nosy neighbors listen for personal enjoyment.

The average radio receiver is specifically designed to receive analog signals and turn them back into sound. This is why eavesdropping is a genuine threat for analog cordless phone and headset users. The frequency modulated (FM) radio signals transmitted by these phones are easily monitored using readily available radio receivers, commonly called "scanners." Thus, our next obvious tip:

TIP ▪ Avoid using analog wireless equipment, and your level of privacy will instantly increase.

Digital signals are much more difficult to decode; these types of transmissions make the decoding process more complicated than with analog signals. In fact, the average person cannot eavesdrop on digital signals without using equipment specifically designed to do so. An additional and very significant level of security can even be added when a digital transmission is encrypted. Depending upon the type of digital transmission and the method of encryption being used, some digital communications devices are more secure than others. None are *absolutely* secure.

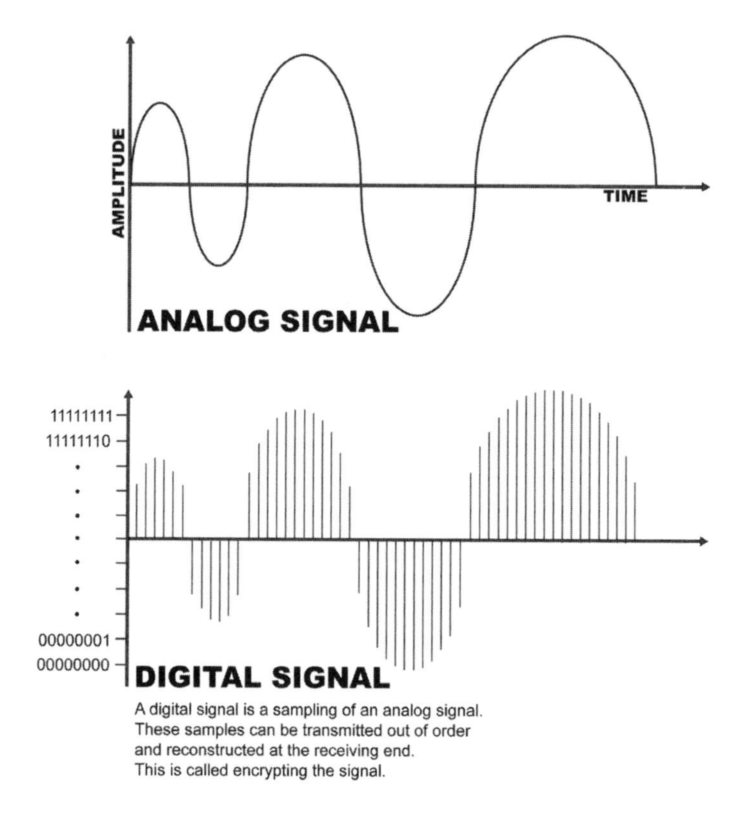

A digital signal is a sampling of an analog signal. These samples can be transmitted out of order and reconstructed at the receiving end. This is called encrypting the signal.

Do not be alarmed by this. The concept of "absolute security" is a myth. Given enough time and resources, commonly used digital communications schemes *can* be decoded. This news may seem disappointing at first. You may be asking, "Do the snoops always have the upper hand? Is communications privacy an elusive dream?" *No.* Let's demystify this and put things into perspective.

Even if your snooper's resources are unlimited (which they *never* are), *time* is on your side. Do you really care if your conversations are eventually decoded? The answer depends on the time period involved:

- **Yes**, if your conversations can be decoded in real time or in just a few minutes.
- **Maybe**, if they can be decoded in a few hours.
- **Probably not**, if decoding takes days, weeks, months, years, or decades. At that point, your conversations become as appetizing as an equally old egg-salad sandwich. Your information is stale—too old to be acted upon.

The truth about security has always been summed up like this:

"How high do we build the wall, my Caesar?"
"High enough to keep them out, you fool!"

Digital communications, especially digitally encrypted communications, keep the eavesdroppers out, giving *you* the real upper hand. At some point, a comfort level is reached and we say, "I feel protected." This is our goal.

TIP ▪ Going wireless? Go digital. Go encrypted digital.

A HISTORY OF CELL PHONES

In the days of 100 percent analog wireless telephone systems (MTS, IMTS and AMPS), generally from 1946 to 2008, eavesdropping was simple. Conversations could be easily overheard by someone simply using a receiver tuned to the correct frequency. Eventually, eavesdropping on these calls became a national pastime, and a law was enacted in the United States. The law prohibited the manufacture and sale of radio receivers capable of eavesdropping on the then-current cellular phone system frequencies. The new receivers were referred to as "blocked" because the cellular frequencies were blocked from being received.

The logic of this law still escapes most thinking minds. Even the manufacturers snickered. They continued, for a while, to manufacture receivers that could be unblocked by the consumers by simply snipping (or adding) an internal wire.

The law did little to stem the eavesdropping. Millions of receivers made before the law went into effect were still being used, and new "unblocked" scanners continued to be offered for sale by mail-order from other countries. Radio shops in Canada and England began doing a thriving business through overnight express services. In a way, it was reminiscent of the bootlegging days of Prohibition.

This ineffective law finally became totally outdated when the cell phone service providers migrated to the digital system we have today. The law itself remains as an odd legal appendix. As hackers find quicker ways to decode digital transmissions, perhaps it will become somewhat relevant again.

Today, cell phones use 100 percent digital transmission. So do many of the home cordless phones and wireless headsets being sold. These digital transmission–based phones and accessories are dramatically less vulnerable to eavesdropping. So in despite of the myth that cell phones are easy to eavesdrop on using a scanner, the reality is, that is no longer true.

When shopping to upgrade your wireless communications devices, you may see digital transmission referred to in the sales literature using one off the following terms:

- www)) CDMA (cell phones)
- www)) GSM/TDMA (cell phones)
- www)) Spread Spectrum/FHSS/DSS (cordless phones)
- www)) DECT (cordless phones)

All these transmission schemes provide adequate security for the average user, if no one has tampered with them and the manufacturers are not being misleading with their claims.

SECURE CORDLESS PHONES: SHOPPING TIPS

A few manufacturers are a little less than forthcoming about their products—shocking, but true. On some cordless phone packaging, "DIGITAL" is printed in big letters on the box in an effort to move old stock. In this case, "digital" refers to some part of the telephone's circuitry being digital. The radio transmission—the part that you, the consumer, are interested in—is still the old, easy-to-eavesdrop-on, FM analog transmission.

Amazingly, at least one major manufacturer even www)) advertises "Digital Spread Spectrum" when only the base station uses this technology; the handset uses the inexpensive, unsecure FM analog transmission. An eavesdropper's receiver, tuned to the handset frequency, easily allows him/her to hear both sides of the call. Unfortunately, you may not be able to tell if the phone uses both technologies just by looking at the box; the manufacturers have made sure of that. Unethical motive or marketing strategy? You decide.

TIP ▪ When products do not specify their transmission type, you should assume they use the worst: FM analog.

TIP ▪ When you purchase a cordless phone, avoid buying the least expensive model. (This group includes FM analog models and the mongrel half-digital, half-analog models described above.)

The price tag is the best clue to avoiding a purchasing mistake. Also, avoid close-outs and specials. These sales gimmicks could be an indication that the seller is unloading old analog or partially digital models. Bottom line: A good digital cordless phone costs more than a similar analog model. The increased level of privacy makes the digital model well worth it. So treat yourself. Buy the best model you can afford. Curiously, the opposite is true of cell phones . . .

TIP ▪ Inexpensive cell phones are preferable because they do not have the capacity to store spyware.

TIP ▪ Cordless phones using Bluetooth digital technology are also available.

Even though Bluetooth is a very good short-range transmission technology, hackers are always working on the next hack attack, so beware. (See the "Bluetooth® Eavesdropping" chapter for additional information.)

By now you may be asking, "So, how can I make sure I buy the most secure cordless phone?" Look for products that clearly indicate either "DECT" or "DECT 6.0." Although this digital transmission standard has been hacked in the lab, it remains secure enough to provide decent privacy for the average person. It also operates on a frequency range different from what computers use for wireless transmissions (802.11x/Wi-Fi). Thus, you'll experience no interference and better call quality.

If you see the terms "Spread Spectrum (SS)," "Digital Spread Spectrum (DSS)," "Frequency Hopping Spread Spectrum (FHSS)," and "5.8 GHz frequency range" on the product label, chances are good that these cordless phones will provide adequate privacy as well, especially among the higher-priced models.

Perhaps you're saying, "Analog? Digital spread? What? I can't even *pronounce* 'GHz.' That's not a word!" Relax. We are just using these terms to be technically correct and uniform in our understanding. Don't worry. You do not need to know the scientific definitions. There will not be a test. Think of these strange words simply as advertising terms. They will become useful when you go shopping and are reading product descriptions. Just match up what you read here with what is printed on the cordless phone's box.

For readers who want to know more, links to all the definitions are on this book's companion Web page at http://www.spybusters.com/Cell911.html.

There is a catch: Eavesdropping on your wireless phone calls is not always dependent upon intercepting the over-the-air signal. "Smart" cell phones, for example, can be secretly forced into sending your conversations, text messages, and more to your eavesdropper without your knowledge. Let's begin to address this in the following chapter.

PRE-BUGGED CELL PHONES

Here is a pop quiz: What is the best way to acquire a new cell phone?

1. Your significant other loves you so much, he/she buys you a beautiful cell phone. Now you can talk, text, send pictures, and even video call each other all the time.
2. You receive an e-mail saying, "You've won!" or "You've been chosen to test . . ." and your new cell phone is on its way, complete with a year's worth of prepaid calling time. "Tell us how you liked it after a year, and it's yours to keep—free!"
3. Your boss gives you a phone, saying, "Use it all you want; we're on a plan."
4. You go to the store, buy the phone yourself, and never let anyone else touch it.

That's right, the best way is No. 4. Here's why: Buying it

yourself is the best way to be sure your new phone has not come pre-bugged with spyware.

www))) Pre-bugged cell phones are new, fresh-out-of-the-box phones that come pre-loaded with spyware. Pre-bugged cell phones are available in almost every make and model and are the *identical* twin of their off-the-shelf counterparts. You will not be able to tell the difference between a pre-bugged cell phone and a normal phone just by looking at them. Both were made in the same factory. They are identical. Only the spyware hidden inside makes them different. This spyware is what makes pre-bugged phones truly "the gift that keeps on giving."

Spyphones were once hard to obtain and were available only in a few models. Times have changed. A recent Google search for "spyphone" returned more than *half a million* URLs—double the number returned for this search term last year (269,000), which at that point was double the number of URLs the year before that (135,000). Today, spyphones are definitely increasingly common out there.

TIP ▪ Do not accept a gift phone from anybody.

Yes, there are exceptions to this tip. Giving a phone to an elderly person or a child for security reasons, for example, may not pose any threat. However, since you were concerned enough to buy this book, this exception does not apply to you. So don't do it.

Spam e-mail sent out by spyphone manufacturers shows great sophistication in marketing. These sales pitches are very informative about the spy features incorporated into the phones they sell. The following is a composite example of the average pitch (provided for educational purposes only):

> Dear Sir,
>
> We are so delighted to present you with our latest innovation, The [name omitted] GSM Spyphone. [Name omitted] utilizes the most powerful covert applications for remotely monitoring an individual's mobile phone activities from anywhere in the world. Our powerful and highly advanced software is the world's first of its kind, and it takes GSM communications to a whole new level. Available on Nokia models: USD$1,650 (with phone)

Main Features:

- Call interception
- Dial in and intercept both sides of target phone conversations
- Immediate text notification when phone places or receives a call
- Text notification shows whether landline, mobile, or international number is calling
- Immediate text notification when the target phone dials an outgoing call
- Text notification shows number being called
- Receive both incoming and outgoing text messages
- Dial in and listen to surrounding vicinity
- Immediate text notification when the target phone is switched on
- Time and date stamping
- Covers all GSM coverage

- Change predefined number on SMS command
- Change stealth mode on SMS command
- Remote activation and deactivation of spyphone feature, text interception feature, interceptor feature, boot/SIM change feature, and text notifications
- Remote SMS command to disable all features
- No logging of forwarded text messages
- No logging of predefined number
- Intelligent auto-delete feature
- Completely hidden software
- Bulletproof reliability
- Future-proof technology

Text Interception with Remote Control:

- Receive duplicate copies of all text messages from target phone
- Covers all GSM coverage
- Receive both incoming and outgoing text messages
- Change predefined number on SMS command
- No logging of forwarded texts
- No logging of predefined number
- Intelligent auto-delete feature
- Remote activation of text interception feature on SMS command
- Remote SMS command to completely uninstall spy software
- Features are completely hidden to phone user
- Bulletproof reliability
- Future-proof technology
- World leaders in spyphone development

Spyphone ("Listening In" Feature):

- Able to secretly dial in and listen to the sounds and conversations coming from the vicinity of the target phone—from anywhere in the world, without the end user knowing.

When faced with such sweeping, privacy-limiting claims, you should be asking just one question . . .

HOW DO THEY GET AWAY WITH THIS?

Good question. In theory, phone companies could analyze their records and identify specific phones that are running spyware. The duplicate e-mails and text messages being sent from the same phone would be a big clue, as would a spike in this activity over the phone's average billing. Once these clues were detected by the software running in their billing programs, the phone companies could alert the affected customers.

Sadly, this is not in your future.

It costs money to create the software, run the programs, and add customer service agents to handle this type of customer relations. And hey, just like other questionable practices—such as charging in minute increments and rounding your time off to the next higher minute—snitching spyphones generate extra revenue for the phone companies! Spyphones make *cents*, and all this makes *sense* to the phone companies.

There is no incentive for the phone companies to do anything about spyphones. In fact, the opposite is true. So you need to rely on *your* street smarts to protect yourself.

SPYPHONE PROTECTION ESSENTIALS

Begin by adding the following tools to your arsenal of protective measures:

- **Never accept a cell phone as a gift.** Period.
- **Make sure your phone isn't swapped with a spyphone.** Mark your current cell phone so that you can identify it as yours. A subtle scratch mark or an invisible UV paint mark should do it.
- **Be suspicious if people tell you your phone was busy when you *know* you were not using it.** Note: Some spyphone models won't be busy when in spy mode due to dual phone numbers on their SIM cards. Your phone's radio transmission will still be detectable, however. You will see the importance of this when using your SpyWarn Mobile™ detector (described in a later chapter).
- Note unusually short battery life. This, plus the phone feeling warmer than usual, may indicate that the phone was being used in eavesdropping mode (assuming your battery is in good health otherwise).

Now that you know better than to accept a gift spyphone, what else do you need to be concerned about? How about a *normal* cell phone being used as a bug? The tips provided in the following chapters address additional ways to protect your privacy, even when your own phone's security has not been breached.

CELL PHONES AS BUGS

It is time for you to play detective. Your first clue is free: A pre-bugged cell phone was *not* involved in either of the following eavesdropping crimes. Your job is to puzzle out how the eavesdropper did it. Hint: The answer is the same in both cases.

Eavesdropping Case No. 1

Your friend complains to you that she thinks her significant other is jealous. Worse yet, she has a feeling that she is being spied upon. References are made to the types of music she plays in her car and, occasionally, to some conversations she has held with other passengers. The most troubling, however, are the bits of cell phone conversations that her significant other repeats to her—in some cases, word for word. The odd part is that they seem to be only the words from your friend's side of the conversation and only from cell phone calls made from the car.

Thinking back, your friend recalls finding an ordinary cell phone under the car seat. Thinking it belonged to one of the kids' friends, she left it on the kitchen counter for the next time the friend came over. Now it's gone. She figured the kid had probably returned for it—right?

Something is wrong. What happened?

Eavesdropping Case No. 2

You and your mate enter an automobile dealership. A pleasant salesman dressed in a jacket and tie shows you the car of your dreams. Wow—you want it! Time to negotiate price. You dive into his cubicle, land on the cushy chairs, and are offered some coffee. Let the games begin.

As you turn the final lap on the bargaining track, the nice salesman excuses himself (bathroom break, talk with manager, etc.), drapes his blazer over his chair, and leaves the two of you alone . . . hoping you will discuss your final price strategy. The size of his sales commission rides on this.

A few minutes after his return, you begin to suspect he is either really good at reading minds, or he heard every word the two of you said while he was gone.

No, he is not a mind reader. If he could do that, he wouldn't be selling cars. He is, however, a car salesman, and some car salesmen are not as ethical as others.

Trust your instincts. You have been had. Either his desk is prewired with a microphone, his desk phone is rigged to be live while the receiver is on the hook, or an ordinary cell phone was left in the pocket of his blazer.

Both of these cases are based on true events. In both cases, electronic eavesdropping was accomplished when the eavesdropper called the hidden cell phone from another phone. The eavesdropper heard every word spoken.

You may be saying, "But I never heard it ring!" Hey, you're bright. You have this halfway figured out already!

Eavesdropping with a normal cell phone is an old trick. It works every time, and it happens all the time, in many situations—from leaving a phone in the boardroom to carrying it along on private business negotiations to stuffing it into a child's backpack before visitation begins.

Let's look at this more closely.

ZOMBIEPHONE: THE NO. 1 CELL PHONE EAVESDROPPING TRICK

An ordinary cell phone by day, it is an evil eavesdropper when forced to the dark side. Even awake, it remains as quiet as the dead. Call it anytime to eavesdrop. It's cheap. It's easy. It works. It's a ZombiePhone. Here's how the eavesdropper does it:

- Activate the Ringer Mute feature so the phone does not make any noise when it is called.
- Activate the Auto-Answer feature so the phone automatically answers whenever it is called.

As long as its battery is being exchanged periodically, the ZombiePhone might be hidden in clothing, a room, or a car. But for long-term spying, ZombiePhone heads for deeper cover. It may lurk inside mains-powered room objects such as a lamp, clock, radio, or television, or it might use a false ceiling or hollow wall as its crypt. As long as it can feed on its AC charger, it is ready 24/7 to suck out the lifeblood of your conversations.

Not all phones have the *immediate* Auto-Answer feature. Some phones will automatically answer only when a headset is plugged in; this is a *bonus*, however, since it allows the microphone to be placed even closer to the conversation. With other phones, some simple Frankenstein-type manipulation is required.

To overcome the lack of an *immediate* Auto-Answer feature, eavesdroppers install a **silent ring tone** or **mosquito ring tone** on the phone. The ring tone is given a plain vanilla name, like *ring* or *default*. Even if someone stumbles across the phone and samples the ring tone during the programming process, he/she will not hear anything. The average person will think, "Hmmm, this doesn't work—maybe it got erased," and move on.

By using this type of ring tone, phones with *normal* Auto-Answer capability (that is, they answer after two to three rings) then have the same eavesdropping capability as the classic ZombiePhone with the *immediate* Auto-Answer feature.

A clever eavesdropper will use a prepaid cell phone for this trick. He/she will pay for it with cash and provide a phony address, if asked, so that even if the ZombiePhone is discovered, it cannot be traced back to the eavesdropper easily, if at all.

This type of bugging device may be difficult for you to find if it has been buried somewhere you can't see it, but do not let that stop you from trying. If you do find a cell phone in bugging mode, remember: All you have is a piece of hardware. Stop and think. You may want to leave it alone, for now, for several reasons. You have not yet tied a suspect to the crime. You have no one to prosecute. Whoever the eavesdropper is, he/she is free to bug again—and you can be sure this person will do a better job of hiding the eavesdropping device next time. A better course of action is required.

WHAT CAN YOU DO ABOUT A ZOMBIEPHONE?

To catch your eavesdropper, you must learn how to test for leaks. You must develop and document a *cause-and-effect relationship* between your information losses and the eavesdropper. You will collect and document *circumstantial evidence* (information that indirectly establishes the facts) of the crime. This is exactly what you need in order to obtain a legal solution to your problem. If you can stomp on your eavesdropper's toes with the weight of the law, it is unlikely that he or she will ever bug you again.

www

"Testing for leaks" means just what you think it means. It's just like a *Mission: Impossible* sting operation: You selectively release one unique, juicy tidbit of information. This forces the eavesdropper to say or do something that he/she would not have said or done had he/she not been eavesdropping. Variations of this technique may be used to ferret out ZombiePhones as well as other types of snoops and eavesdropping devices.

This simple testing process will do several things for you. First, it will establish whether or not there truly is an active eavesdropping problem. It can also help narrow your search to a specific area (a particular room, vehicle, or person), thus giving you a better chance of finding the hidden ZombiePhone.

The specifics of how to conduct your tests are explained later, in the "Murray's Test for Leaks™ Protocol" chapter. Once you have performed these tests and documented your circumstantial evidence, consult with an attorney who specializes in digital technology to decide what your next step should be.

Oh, and that missing cell phone from the kitchen counter? Nobody seems to know anything about it, although the significant other is "quite certain" one of the kids' friends just saw it there and took it back.

Sure.

Now, to some of you, this might sound like a worst-case scenario. Right now you might be saying, "What could be worse than a ZombiePhone?" You will find out next what can be worse: GSM bugs.

GSM MICRO-BUGS

A cousin to the ZombiePhone www)) is the GSM micro-bug. These are miniaturized cell phones made specifically for covert eavesdropping! Like ZombiePhone bugs but without normal cell phone features, these are tiny, creepy, robotic, cell phone bugs often hidden in such everyday objects as power strips and lighting fixtures. Their tiny size is possible because they do not have keypads, ringers, displays, or smart-phone features. When called from *any* other phone, they become eavesdropping bugs *automatically.*

Groupe Spécial Mobile (GSM) is the name of the world's most popular cellular telephone standard. GSM micro-bugs work on this standard, which means they can work in almost anywhere on Earth where there is cellular telephone service. Like normal cell phones whose features are set to Auto-Answer and No Ring, GSM bugs are equally hard to detect because they sleep most of the time. The thing that awakens them is the call from the eavesdropper. Some models also awaken when they hear sound being made near them. Should *you* awaken one, it will silently call the eavesdropper.

If you feel you are being eavesdropped on and you are sure *your* cell phone is free of spyware, a GSM bug may be the culprit.

Bug microphones are much more sensitive than most people realize. Ideally, bugs are placed as close to the sound source as possible, but the rule of thumb when searching is: *If your ear can hear it, so can the bug.* The microphones in GSM mirco-bugs are very sensitive and can capture sound from large areas like bedrooms, offices, and vehicles.

TIP ▪ When searching for bugging devices, you must look in any place where your conversation can be heard.

In a residential setting, if living room conversations are the target, search that room as well as every other part of the house where living room conversations may be heard. An eavesdropper may have decided to place the bug in the home's most central location so that sounds anywhere in the house can be heard equally well. Of course, never assume there is only one bug.

Some GSM micro-bugs are battery operated and look like small, black plastic boxes. Most of them are very small, approximately 2 inches x 1.5 inches x 0.75 inches (55 mm x 40 mm x 20 mm). The smallest one seen to date measures a minuscule 1.18 inches x 1.18 inches x 0.59 inches (30 mm x 30 mm x 15 mm). GSM bugs are also sold pre-hidden in such everyday objects as power strips, lamps, clocks, radios, cordless phones, computer mice, car rearview mirrors, USB memory sticks, and even children's toys! (Note to parents with visitation rights: Beware of the teddy bear!)

Internet spy shops advertise GSM micro-bugs. Many www)) sell for less than $25 on eBay. The standard, broken-English advertising pitch goes something like this:

> This device allow you spy anyone in any place. You only need buy new SIM card, place it in this device. Once you call phone number, this device will auto-answer. It work just like cell phone with auto-answer, but without the keypad, ringer, and speaker. You can hear actually if anyone is talking about you, or whatever, when you call the SIM card's phone number. Features a unique super-slim design. Place it at home or in office. You could listen to conversations wherever you want. Please do not use it to engage in any illegal activity.

Some types of GSM micro-bugs also have insidious second, third, and even fourth creepy features. These are the four main types, along with their worrisome features:

- **Basic Eavesdropping Model.** Activates automatically upon receiving a phone call from the eavesdropper.
- **Advanced Eavesdropping Model.** In addition to the basic Auto-Answer feature, this model will also Autocall when it hears sounds nearby. Most often, before hiding the device, the eavesdropper will preprogram the number it should dial. Some models allow remote programming of this number by making an incoming phone call. Some models also have the ability to call multiple numbers if the first one they try does not answer.
- **Audio/Photo Model.** In addition to the previously mentioned features, this model can be activated by motion in the area. Upon awakening, it calls the eavesdropper and sends audio while a built-in camera sends photos.
- **Audio/Photo and GPS Model.** This model has a built-in GPS that sends tracking information in addition to audio and photos. The eavesdropper can map the location, time, date, and speed on his/her computer while listening to live conversations and seeing photos taken at regular intervals. Other GPS models exist with lesser capabilities. Two things are common to all models in this group: the GPS tracking ability and the use of the cellular phone system

to transmit data. If the capability is solely GPS tracking, then it is simply called a "GPS tracker" and it does not have *real-time* tracking ability.

Bugs within this last category—the GPS models—are being used *overtly* by parents who loan their car to their teenagers. Teens hate it! They are also being used *covertly* to track significant others. When these partners find out, they hate it too.

How do these teenagers and partners—and you—discover that they are being bugged by a GSM model?

((www

TIP ▪ If you hear raspy, chattering noises from your audio system at home, at work, or in the car, it might be a GSM bug or a hidden GSM mobile phone placed somewhere near the audio equipment or wiring.

To help you recognize this noise, an audio sample appears on the book's companion Web page at http://www.spybusters.com/Cell911.html.

Not all cell phones use GSM as their transmission mode, however, especially in the United States. Here, we also use a digital technology called Code Division Multiple Access (CDMA). Now, you may be asking, "Can I spot CDMA cell phones the same way?" The answer is typically *no*. These phones do not generate the noise mentioned in the last tip. The lack of the GSM transmission noise around amplified speakers makes CDMA cell phones less prone to accidental discovery. Techno-savvy eavesdroppers might find this an attractive reason to use a CDMA phone as their bugging device, despite the fact that its larger size makes it more difficult to hide.

The good news is, due to global economics, CDMA is rarely used to make this type of bug—hence the name *GSM* micro-bug. Both CDMA *and* GSM cell phones are, however, being used as ZombiePhones. Both are also capable of having spyware loaded onto them.

TIP ▪ It is easy to tell the difference between GSM and CDMA phones and bugs. GSM phones and bugs have SIM cards; CDMA phones and bugs do not.

At this point, you know the secrets of phones pre-loaded with spyware, normal cell phones hijacked for covert bugging operations, and insidious GSM bugs that use the cell phone system as their own surveillance playground. Just being aware that these vulnerabilities exist makes it easier for you to defend yourself against them. The possible methods for invading your privacy are narrowing.

The big question, however, still remains: "Is my cell phone bugged?"

WHAT IS CELL PHONE SPYWARE?

The questions I hear invariably start with two words: "Can someone . . . ?"

They continue with the following inquiries:

". . . listen in on my calls?"

". . . listen to my voice mail messages?"

". . . remotely steal my contacts list?"

". . . send fake texts from my phone?"

". . . activate my microphone 24/7?"

". . . make my phone dial someone else?"

". . . get a text stating the length of my call?"

". . . get a text when I use my phone?"

". . . send me texts using a fake number?"

". . . get my new phone number when I switch SIM cards?"

". . . get a text with the numbers I call and receive?"

". . . track where I am using the phone's GPS?"

". . . track where I am even if my phone lacks GPS?"

". . . record my calls using my phone's own memory?"

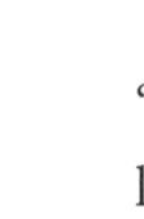

"... trick me into installing spyware by making it look like a game?"

"... do all this from anywhere in the world?"

The next question I hear is, "Isn't this illegal to do in the United States?"

The answer to all these questions is *yes*. Someone can do any or all of these things using spyware, although most of them are illegal in this country. (Generally speaking, the law is usually about ten years behind the technology.)

Spyware is part of the larger ((www malware family of mini com- ((www puter programs. It is also called a "Trojan" (as in "Trojan horse") because to get past security measures and into your phone, it masquerades as something more benign. Spyware allows unauthorized people to secretly spy on the activities of the computer system, cell phone, or tablet device it has infected. You may also hear spyware referred to as a "computer virus," but to be accurate, viruses are into asexual replication (that is, they attempt to reproduce themselves and spread to other devices). Trojans simply invade, stay put, and do their job.

Mobile spyware, as the name implies, is specific to portable computing devices. Many cell phones and tablet devices now have computer abilities, so they are commonly called "smart"—and just like computers, they have become vulnerable to spyware infections. The smarter the phone, the more things spyware can do. Because spyware resides on devices that communicate using the cellular telephone system, spies have the freedom to monitor from anywhere in the world.

HOW DO CELL PHONES BECOME INFECTED?

There are several clues that can tell you whether your cell phone has been infected; we will address these more specifically in the next chapter. But keep in mind: One clue alone does not automatically indicate a spyware problem. A battery that runs out more quickly than it used to, for example, could simply be a faulty battery or one reaching the end of its life span. The more clues you notice, however, the more likely it is that your phone has a spyware problem.

Most often, spyware programs must be *manually* loaded into your phone's memory; someone can do this fairly quickly. Direct installation, the most common way spyware gets into phones, can be accomplished in just a few minutes. This is why it is *so* important to password-protect your phone and maintain physical control of it at all times.

Spyware programs can also masquerade as games, wallpaper, ring tones, or an electronic business card (vCard) ((www specifically designed to trick you into allowing them onto your phone. You may not have a clue that you have fallen for the trick until you notice that something is not working correctly.

TIP ▪ If your newly loaded game, ring tone, or other doodad does not work properly, or if your phone begins to act sluggishly, you may have inadvertently installed spyware.

Sometimes spyware is sent to a phone as an e-mail attachment; often it is made to look like a game or other application. This tricks the phone's owner into running the program, but what you are really running is a spyware installer. Do not open e-mail attachments if you are not positive they are safe. They could be a spyware Trojan horse.

Sometimes merchants of privacy mayhem buy new phones and load spyware onto them. They then sell these pre-loaded spyware phones to the general public. Sales of these phones are mainly conducted via spy shop websites.

TIP ▪ Mark your phone so you will know if it has been swapped with a spyphone of the same type. (((www

For a quick education about cell phone spyware and how it works, watch one of the many videos hosted by YouTube.com or visit some of the many websites selling spyware. (((www

SPYWARE PREVENTION CHECKLIST

Below is a checklist of tips for keeping spyware off your phone. While not every tip is practical for every individual, the more tips you can use, the less privacy you will lose. A healthy dose of paranoia goes a long way toward protecting your phone—and your privacy.

- Do not accept a phone given as a gift.
- Never loan your phone to someone else.
- Never let someone install a free ring tone, wallpaper, a "cool app," or an e-mail attachment they just sent you.
- Never let your phone out of your possession, ever.
- Do not jailbreak your phone.[4]
- Lock your phone with a password. Keep it private. Change it occasionally.
- Keep your phone in your possession at all times.
- Do not use your old SIM card in your new phone. Conversely, do not put a new SIM card into your old phone. (Some spyware has the capability of detecting a SIM card swap and will immediately report your new phone number to the spy.)
- Do not download an app, wallpaper, ring tones, etc., offered in an unsolicited text message or e-mail. (If you want this doodad, obtain it from the website after verifying that the provider is legitimate.)
- Do not let your "friends" download anything for you either.
- Use the most restrictive of your phone's settings for apps and Internet access. (Some phones will even flag the activity and warn you if the program tries to do more than it has been given permission to do.)
- Limit the number of apps you download to the essentials. (Spyware has been known to masquerade as a "fun app." The program *Tap Snake*, for

---

4 According to http://en.wikipedia.org/wiki/IOS_jailbreaking, "jailbreaking" is a process that allows iPad, iPhone, and iPod Touch users to run third-party, unsigned code on their devices by unlocking the operating system and allowing the user root access

example, is an Android app that reports the phone's GPS location coordinates to the spy once every 15 minutes. Downloading a spyware app is an easy mistake to make—more so on some mobile operating systems than others, due to varying degrees of oversight by the system developers. Not all apps are screened by the phone carriers or manufacturers for spyware.)

- Use an inexpensive phone; it will not have the capacity to hold spyware. www))
- www)) Do not accept a gift phone. (Remember, a Trojan horse is also known as a "gift horse.")
- Use a CDMA-type phone if possible. (Most current spyware is written for GSM phones. This may change, but until it does, using a CDMA-type phone may reduce your risk.)
- Follow the advice in the "Murray's Test for Leaks™ Protocol" chapter. (Periodically make calls and send texts that will evoke a response from potential eavesdroppers, thus forcing them to reveal themselves. In doing so, do not let it become generally known that you are suspicious.)
- Keep your phone turned off as much as possible.
- Change phones, SIM cards, and carriers once in a while. (If your employer supplies your phone, this may not be practical.)
- Consider using a prepaid phone.
- www)) Keep up with the news on mobile spyware at sites like http://threatcenter.smobilesystems.com.

"What if they've already nailed me?"

> **TIP** ▪ New spyware is constantly emerging. Computer security researchers report finding bogus Android apps (mostly games and wallpaper) that mimic real apps but contain Trojan horse code. Stay away from the alternate-apps marketplaces when purchasing the apps you need.

If you have been infected with spyware, the cheapest solution is to reinstall your phone's operating software. Some phones, like iPhones and BlackBerrys, allow you to do this yourself. If you are not sure how to do this, contact the phone manufacturer or your place of purchase. See "Reinstalling cell phone software" at http://www.spybusters.com/Cell911.html for further information.

The easiest solution, of course, is to buy a new phone. Do not reinstall your data from a backup file; it may be infected with the spyware. Delete your old backup file. Start from scratch. And this time, follow the Spyware Prevention Checklist.

APPS AGAINST TAPS

Battery-monitoring applications can help spot mobile phone spyware in your phone, especially the kind that enables live eavesdropping. Some apps offer graphs showing battery life over a time period. Just after the phone is charged, the graph shows 98 percent to 100 percent charge.

It declines gradually during the day, with big percentage drops when you use the phone. If you see big drops during periods when you know you were not using the phone, this power drain would be very suspicious, indeed. You might suspect spyware at work, assuming your battery is in good condition otherwise.

There are dozens of battery-monitoring apps.[5] Evaluate several of them before purchasing one; some of them do not have a graphing function. Also, remember that the more automatic features of your other apps you can turn off, the easier it will be to spot spyware in operation. Did you know there are even apps for turning your phone's features off and on if you find it difficult to use your control panel? ((www

www)) Spyware protection applications can help protect you against *accidental* downloading of spyware and other malware onto your phone. These software programs and services do not rely on examining your phone's software. Instead, they block spyware *before* it gets to your phone, thus preventing you from pressing that "Install" button.

Spyware protection apps also have several other useful security features, such as automatic backup and storage of your data and the ability to track your phone and erase your sensitive data if it is stolen. Granted, depending on the nature of the operating system architecture and features, your phone manufacturer may provide these services to you by default. But add-on security software is useful to people with other operating systems such as Android, Windows Mobile, and BlackBerry.

---

5 Due to the ever-changing world of apps, making specific app recommendations is not practical here. Instead, conduct an Internet search to locate the latest and greatest applications. Battery-monitoring apps for any mobile phone operating system may be found using the search terms "battery app" + "[your phone or operating system]."

All the major security software vendors—Symantec, F-Secure, Kaspersky—have some type of cell phone security offering. Several small start-up companies—Lookout, Inc., is an example—are also into the counterspy game, offering some impressive protection applications. Android phone users are particularly fortunate. The DroidSecurity™ program is available to them for free.

TIP ▪ Use an Android cell phone to get built-in security—for free.

Billed as "the first full-featured consumer anti-malware and physical security app for Google's Android operating system," the DroidSecurity™ suite of protections now has more than 2.5 million users.

"The tips are helpful, but is there a way I can determine if my cell phone is infected with spyware?"

Good question. Your answer is in the next chapter.

SPYWARE DETECTION: THE MURRAY METHOD™

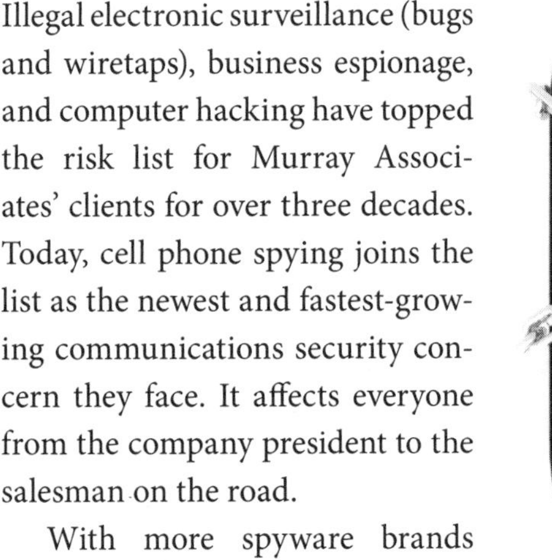

Illegal electronic surveillance (bugs and wiretaps), business espionage, and computer hacking have topped the risk list for Murray Associates' clients for over three decades. Today, cell phone spying joins the list as the newest and fastest-growing communications security concern they face. It affects everyone from the company president to the salesman on the road.

With more spyware brands becoming available via the Internet, more individuals purchasing high-end phones, and the subsequent drop in prices of both, spyware is now a pandemic privacy issue. *Everyone* who uses a smart cell phone is affected.

One of the hallmarks of spyware is that it is very stealthy. Even an expensive, full forensic examination cannot guarantee that well-hidden spyware will be discovered. This news, of course, does not please corporate presidents. A better answer was needed. A quick and sure method of detecting an infected phone had to be found.

What developed from talking with our clients was a methodology specifically designed to fulfill the following requirements:

- It must make quick and reasonable spyware evaluations.
- No special forensic tools should be required.
- No special skills should be necessary.
- No assistance should be necessary once the initial training is over. The phone owner must be able to conduct the test him- or herself—anytime, anyplace.
- Advancements in spyware software and cell phone hardware should not render the test ineffective.

The field of cell phone forensics is new, growing, and changing at a rapid rate. Better methods of detecting spyware may come along. Cell phones may evolve to become immune to infections. New surveillance tools will probably develop along with new communications methods. If so, counterespionage practitioners will keep pace, as we always have since the beginning of time.

In other words, keep an eye out for the next edition of this book.

Currently, however, the Murray Method™ is state of the

art. It fulfills our corporate and government client needs, and it will work for you, too.

Making a determination that your phone is bugged with spyware is mostly a matter of observation. Murray Associates has identified nineteen warning clues. All of these clues are easy to observe. The more clues you see, the more likely there is a problem. The final determination is made by combining your clue count with your phone's **Duty Cycle Ratio**, a measurement that we will explore later in this chapter. This measurement tells you how often the phone spends its time transmitting when it is not actively being used by you.

Before we start counting clues, you need to keep in mind that some clues may also reflect issues *other* than spyware—your battery dying prematurely simply because it has reached the end of its life span, for example. However, the more warning clues you notice, the more likely your phone *is* infected. Put a check mark next to any of the following suspicious behaviors you have noticed.

YOUR PHONE IS INFECTED: THE 19 WARNING SIGNS

Many of the following clues have one thing in common: They indicate that your phone is transmitting when you are not using it.

1. Your phone's battery life suddenly decreases.
2. Your phone periodically lights up for no apparent reason.
3. Your phone is unusually warm when you are not using it.

4. Your phone bill shows a spike in SMS (short message service), text, or data usage.
5. Odd background noises or clicking can be heard during calls.
6. Friends *often* say your phone was busy or jumped straight to voice mail, when you know *you* were not using it.
7. Your phone *beeps* for no apparent reason.
8. Your Call Duration log shows entries that don't make sense to you.
9. Features of your phone that you didn't activate, such as call forwarding, are *on*.
10. Your phone frequently displays error messages such as "App Closed: Main" (or similar).
11. Your phone frequently displays "Message stuck in outbox" errors.
12. Your phone is slow to respond to keypad entries.
13. "No SMS credit" messages appear on the screen.
14. Your phone is receiving odd text messages (example: <^#83><2125551212><d>).
15. The Web icon appears for no apparent reason.
16. You notice weird additions in your phone's Applications menu.
17. Your phone does not turn off as quickly as it once did.
18. Specific details of your calls are being mentioned to you by others.
19. Your phone shows frequent phone transmissions when you are not using your phone.

Detecting an increase in transmissions is the biggest clue that your cell phone is infected with spyware. The phone must transmit when it sends information back to the spy: your texts, your e-mails; data about who you called, when, and for how long. It will also be transmitting when your spy has called your phone and has silently turned on the microphone to listen to what you are doing.

So how can you get this information about your phone's actual transmissions? By using a SpyWarn Mobile™ cell phone transmission detector.

THE SPYWARN MOBILE™ TRANSMISSION DETECTOR

Cell phones have secret lives. When we are not using them, they chitchat with their masters at cell phone central, periodically transmitting messages like, "I'm here! I'm ready to accept a call." They also download e-mail and news, adjust their transmitter power settings, and switch cell sites automatically. Depending upon the make, model, and type of apps loaded onto the phone, the phone may be handling other chores as well. In short, your phone transmits when you are not using it, and this is perfectly normal.

A number of other types of transmissions, however, are not normal. They may indicate that someone is receiving private information directly from your phone. And employing a SpyWarn Mobile™ cell phone transmission detector is the quickest and easiest way to check for these abnormal transmissions.

A SpyWarn Mobile™ device is available—FREE—to everyone who buys this book new, either printed or as an e-book.

- If the SpyWarn Mobile™ was not included with the book you bought, you may request it by mail. Simply fill out the coupon on the last page of this book and send it in. Only original coupons can be honored, for obvious reasons.
- If this is an e-book, please e-mail a copy of your purchase receipt along with your mailing address to SpyWarn@spybusters.com.
- If this is a *used* book and the coupon is missing, you may still obtain one or more SpyWarn Mobile™ devices using PayPal, at www.spybusters.com/spywarn.html.

SpyWarn Mobile™ is guaranteed for 90 days. A free replacement is offered during this time period should a non-battery failure occur. Simply send us your old detector, along with your name and address, and we will promptly send you a replacement.

HOW SPYWARN MOBILE™ WORKS

Spyware works by forcing your phone to transmit your information to the person who is spying on you. These transmissions usually occur when you are not using the phone. There is one exception: when the spyware is eavesdropping on an *active* call. However, if your spy is doing that, he/she is also transmitting information when you are not using the phone, and SpyWarn Mobile™ will still help you.

SpyWarn Mobile™ lets you *see* when your phone is transmitting. Just place it near your phone; within 1 to 4 inches is a good distance. It will begin blinking when it senses your phone is transmitting. SpyWarn Mobile™ works particularly well in sensing GSM phone transmissions, the technology favored by most spyware.

Along with the other warning signs of spyware discussed in the previous chapter, seeing unusual transmission patterns provides you with one more clue—a major, irrefutable clue—that spyware is present on your phone.

Setting up your SpyWarn Mobile™

To begin using SpyWarn Mobile™, insert the battery. Place the battery, *plus (+) side down*, into the rear of the SpyWarn Mobile™. Then slip the battery partially into the unit. SpyWarn Mobile™ will blink briefly to let you know it is working.

You may leave the battery in place, as there is very little drain on the battery when it is not blinking. The battery will last long after you have finished testing for spyware; it is replaceable with any of the following types of batteries: CR1220, DL1220, ECR1220, BR1220, or SB-T13. Batteries must meet the following specifications: 3 Vdc, 36 mAh, Lithium, 12.5mm x 2.0mm.

THE SPYWARN MOBILE™ TEST

Determining your phone's transmitting time is easy to do using the accompanying SpyWarn Mobile™ device. It will show you, by flashing, when your phone is silently transmitting. Simply measure the time your phone spends transmitting when you are not personally using it to make calls, send text messages, or download e-mails. Using this measurement, you will determine your phone's **Duty Cycle Ratio**; a percentage number that can reveal if spyware is at work.

TIP ▪ Even if you don't have your SpyWarn Mobile™ detector, you may start testing immediately by placing your cell phone (GSM types only) near a radio or other piece of audio equipment. You will hear a raspy buzzing noise[6] when a transmission occurs. Note the start and stop times of this noise, and apply them to the formulas below.

TIP ▪ The longer the duration of your phone's transmissions, the more likely spyware is at work.

A small amount of buzzing is normal when the phone "shakes hands" with the nearest cell tower or downloads e-mail. More constant buzzing, along with an unusual drop in battery power, is a good indication that spyware is on your phone and that someone is using it as an eavesdropping device. The secret to detecting abnormal activity lies in the *comparison*.

- How often is the phone transmitting?
- How long are the transmissions?
- How does this information compare to transmissions from a similar model phone that is believed to be spyware-free?

---

6 You can hear what this buzzing sounds like at http://www.spybusters.com/Cell911.html.

- How does the same phone compare at different times of the day/week, when spyware is most likely to be active vs. inactive?

Before we conduct actual tests using the SpyWarn Mobile™ let's review what you have learned so far.

Measuring transmission time

You already know some key points about evaluating your cell phone for spyware. You know the nineteen signs that your phone is infected. You know that the more clues apply to you, the more likely your phone is infected with spyware. You know that when your phone transmits for longer than a minute (when you are not using it), this is a very strong indicator of eavesdropping via spyware, especially if it happens often *and* you did not just receive several long e-mails with attachments. Your observations so far give you plenty of information with which to make a fair judgment. To make a more accurate judgment and be a bit more scientific about your evaluation, map your phone's **Duty Cycle Percentage** or its **Duty Cycle Ratio**.

- **Duty Cycle Percentage (DCP).** The percentage of time your cell phone transmits when it is not actively being used by you. This figure will normally be around 5 percent, based on the amount of time the SpyWarn Mobile™ is seen flashing. The lower the DCP number, the better. A significantly higher number indicates that your phone is transmitting much more often than the average phone—a strong indication spyware is active on your phone.

- **Duty Cycle Ratio (DCR).** The ratio of *standby time* to *transmit time* when you are not actively using your cell phone. This ratio should be somewhere around 20:1 (95 percent standby to 5 percent transmit). The higher the DCR ratio, the better. A much lower ratio indicates that your cell phone is transmitting more often than the average phone; again, this is a strong indicator of active spyware on your cell phone.

The DCP and DCR figures are directly related to each other. They simply present two different ways of bringing your test measurements into focus for a meaningful evaluation. This may sound difficult, but it is really a fairly easy and very worthwhile exercise.

Again, you start by simply making note of the time the SpyWarn Mobile™ starts flashing and stops flashing. Adding and subtracting minutes and seconds is the only tricky part, but there are calculators on the Internet that can do it for you. ((www

Important evaluation tips

When conducting your SpyWarn Mobile™ test be sure there are no other transmitters in the immediate area. Wireless items—such as cordless headset or phone, another cell phone, keyboard, or mouse—or a nearby Wi-Fi station create radio-frequency energy which may also cause the SpyWarn Mobile™ to blink. Test the area for fase positives by turning on and observing the SpyWarn Mobile™ before bringing your cell phone near it.

Remember, normal phone activities will cause SpyWarn Mobile™ to flash. These include your e-mail and text transmissions (incoming and outgoing); the moment just before the phone rings with an incoming call; during phone conversations; and periodically, when your phone handshakes with the cell tower. SpyWarn Mobile™ will also flash when you insert the battery, and sometimes handling it will even start it flashing (your electro-magnetic personality, perhaps?). All of this is normal. You will come to recognize these transmissions for what they are rather quickly. What you wish to look for and measure is the flashing that happens when you are *not* participating in any of these activities.

Also, note that spying activity is more easily and *positively* detected by temporarily turning off some of your phone's features during your test. In your phone's setup menu, turn off such features as automatic "fetch/push/notification" services, Wi-Fi, Bluetooth, and 3G or 4G. Options for disabling some services (3G/4G, for example) may not appear on some carrier-branded device menus. If you cannot turn off 3G or 4G, rely more heavily on the nineteen clues mentioned earlier in this chapter to make your determination. If you need help turning your phone's features off or on, just visit your favorite search engine and enter search terms like these:

- (For 2/3/4G) "_G on off" + "[your phone or OS]"
- (For WiFi) "wifi on off" + "[your phone or OS]"
- (For Bluetooth) "Bluetooth on off" + "[your OS]"
- (For Auto sync) "auto sync" + "[your phone or OS]"
- (For GPS) "GPS on off" + "[your phone or OS]"

Be aware that SpyWarn Mobile™ will occasionally blink briefly (approximately 5 to 10 seconds at a time) when you are not using the phone. It is likely that your phone is automatically registering with a nearby cell tower during these times. This can occur once every 10 to 30 minutes. The SpyWarn Mobile™ device will also blink for longer periods (usually 12 to 40 seconds) when your phone is automatically retrieving e-mail or when information is "pushed" to your phone automatically by certain apps. Temporarily turning off applications that use *automatic* services makes the *anomalous* spyware transmissions easier to spot.

Truly suspicious behavior includes slightly longer blinking activity—for example, just after every phone call or multiple of calls (perhaps every fifth call, as information is stored and sent later to reduce the possibility of detection). This indicates that the spyware is reporting data about your recent phone usage or forwarding copies of your e-mails and text messages elsewhere. Since you will not be using your phone during the Duty Cycle test, you may not see this activity. You can conduct additional tests using SpyWarn Mobile™ to determine how soon your phone turns off after completing calls, sending text messages, or transmitting e-mails. The longer it takes, the more likely it is that spyware is at work.

Some spyware packages allow eavesdroppers to call your phone and silently activate its microphone. Extended periods of flashing—especially periods that last *longer than a minute*—when you are *not* using your phone means your phone may be in eavesdropper mode, with someone *actively* listening to you! *Be very suspicious if you see this type of activity. Be careful what you say. Someone is listening.* It is this particular feature of spyware that has the greatest effect on the life of your battery—and why a phone infected with spyware often feels warm to the touch when the owner has not been using it.

The ultimate goal is to determine via testing and observation how hard your phone's transmitter is working for someone else.

Duty Cycle test directions

Set aside 1 hour to observe your phone's transmissions. It should be an hour when you are not using your phone and are not expecting any calls. Start by placing your phone on a table where you can keep an eye on it.

Place a clock (with a seconds hand or digital readout) nearby, or if you are sitting at a computer, you may conduct the test directly online at http://www.spybusters.com/Cell911.html by simply pressing your keyboard space bar to turn our **SpyWarn Mobile™ Timer** on and off.

Place SpyWarn Mobile™ *near* your cell phone.[7] Resist the urge to move either the SpyWarn Mobile™ or your phone during your test.

It is a good idea to allow some sounds in the background while you are conducting your observations. This will help keep an eavesdropper on the line if he/she calls during your test. Playing the radio or TV is acceptable, but if your eavesdropper is cautious, you might want to go the extra step of playing a recording of conversational babble or ambiance www))) sounds. Using sound backdrops will keep the eavesdropper intrigued enough to stay connected to your phone. The

7 Placing SpyWarn Mobile™ on top of the phone is not recommended (1 to 4 inches away works best for most phones).

longer he/she keeps the phone transmitting, the easier it is for you to detect eavesdropping is taking place. Several ambiance sound clips are located on this book's companion Web page (www.spybusters.com/Cell911.html) for you to use. The clip list includes people talking, transportation sounds, and even eerie sounds broadcast by the planet Saturn (courtesy of NASA). You can custom make your own clips from places you frequent, or you can use free Internet sound samples or even the aptly named Alibi CD. ((www ((www ((www

There are two ways to compile your findings: using the manual timing method or using the online SpyWarn Mobile™ Timer, which simplifies the process greatly. ((www

The manual method

Make a simple, three-column chart on a piece of paper to keep track of when SpyWarn Mobile™ flashes.

- The first column is for the Start Time (ST) of the flashing.
- The second column is for the End Time (ET) of the flashing.
- The third column is the actual Time Flashing (TF) in seconds.

Watch for the flashing to begin. This will mark the start of your hourlong observation. Jot down the Start Time (ST) and End Time (ET) of the flashing as they occur. At the conclusion of your observations, compute the Time Flashing (TF) in the last column.

The flashing time will rarely be longer than a minute, but remember to do a minutes-to-seconds conversion if necessary.

A test which was started at 1:52 PM (and seventeen seconds) might look like this, with only the actual minutes and seconds (mm:ss) being shown for clarity:

ST	ET	TF
52:00	52:09	00:09
52:40	52:44	00:04
05:43	06:05	00:22
10:21	11:02	00:41
14:37	14:41	00:04
22:42	22:50	00:08
33:43	33:53	00:10
40:31	41:02	00:31
44:26	44:32	00:06
51:54	52:00	00:06
Total Time Flashing		141 seconds

The formula for measuring the Duty Cycle Percentage looks like this:

Duty Cycle Percentage = TF / (3,600[8] – TF) =
_____ x 100 = _____ %

So in this example, the math looks like this:

Duty Cycle Percentage = 141 / (3,600 – 141) =
.04 x 100 = 4%

The lower the DCP, the better. This cell phone user's DCP is low enough that there should be no cause for alarm,

8 Note that 3,600 = 1 hour's worth of seconds. For simplicity, we use this figure in both the manual and the online formulas. Your test hour may be slightly shorter or longer. Feel free to use the exact number of seconds if you wish, but if you are within (plus or minus) two minutes of a full hour, then 3,600 seconds may be reliably used.

assuming that a majority of the other warning clues are not present either.

You determine the ratio of the Duty Cycle using the following formula:

Duty Cycle Ratio = 100 / Duty Cycle Percentage

So for the example, determine the Duty Cycle Ratio as follows:

Duty Cycle Ratio = 100 / 4.0 = 25:1

The higher the DCR, the better. This cell phone user's DCR is low enough that there should be no cause for alarm, assuming that a majority of the other warning clues are not present either.

The online SpyWarn Mobile™ timing method

The online timer at www.spybusters.com/Cell911.html automates the timing process described above. Here is how it works: Each time you press your keyboard's space bar, you record *both* the total time flashing *and* the total time not flashing. This makes the test very easy to conduct.

Start your hourlong observations by pressing the space bar when SpyWarn Mobile™ first starts flashing. Then, press the space bar again when it stops. Continue to press the space bar once every time the flashing starts and again when it stops. Run the test for an hour. Don't fall asleep.

Using this method, you might create a chart spanning about an hour that looks like this, with the time being expressed using the *actual* minutes and seconds (mm:ss):

ON (time flashing)	OFF (not flashing)
02:21	57:39

In this example, your math for converting Time Flashing (TF) to seconds looks like this :

mm x 60 + ss = TF (in seconds)
2 x 60 + 21 = 141 seconds

Duty Cycle Percentage = TF / (3600 - TF) =
_____ x 100 = _____ %
Duty Cycle Percentage = 141 / (3600 – 141) =
.04 x 100 = 4%

Duty Cycle Ratio = 100 / Duty Cycle Percentage
(The lower the DCP, the better.)
Duty Cycle Ratio = 100 / 4.0 = 25:1
(The higher the DCR, the better.)

While carrying out this test, one thing you will discover is that your phone keeps chattering even when you are not using it. This is normal. The real question is, *how much* is your phone chattering? Most phones chatter in the background about 5 percent of the time. A Duty Cycle Percentage in the *5 percent range or less* is good. The higher your Duty Cycle Percentage is over 5 percent, the more likely spyware is at work.

Conversely, a Duty Cycle Ratio in the *20:1 range or higher* is good. The lower your Duty Cycle Ratio is, the more likely spyware is at work.

In addition to determining your cell phone's duty cycle, there are other revealing clues hidden in your experiment.

Look at the Time Flashing (TF) seconds that you computed manually. Most flashes are of short duration, almost always less than 45 seconds, and sometimes as brief as 4 to 6 seconds. Someone who is eavesdropping while you are not using your cell phone will likely listen for much longer than that, *especially* if something interesting is going on. Be sure to bait this hook with something irresistible. Depending upon the circumstances, this could be anything from business chatter to the sounds of love.

If you compute the Time Between Flashing (TBF), you will notice that flashes occur at fairly regular intervals—either one right after the other, or between 4 and 12 minutes apart. Paying attention to this extra information will give you a feel for what is normal, especially when a comparison is made to a similar phone that is not believed to contain spyware.

The sample charts shown here were made using an iPhone (without spyware) on the AT&T network. A chart from a different phone, on a different network, may look somewhat different. For the best comparison, also conduct a Duty Cycle test with a phone that is similar to yours, with the same features enabled, believed to be spyware free, and using the same network. In each test, take readings over a period of at least an hour. This is the best baseline for comparison.

Transmissions after you use your phone

Spyware also sends phone, text, and e-mail information to the spy once a person *stops* using their phone. In some cases, this data is stored and transmitted periodically—for example, at the end of every fifth call. Look for this activity. It is another important clue you will use when deciding whether your cell phone is bugged.

Use the SpyWarn Mobile™ device to conduct these additional tests. (Remember to turn off your phone's Wi-Fi, 3G/4G and push services first. Place your phone on a table with SpyWarn Mobile™ 1 to 4 inches away from it.)

- Send a short e-mail. You may notice that SpyWarn Mobile™ starts flashing when you access the e-mail feature on your phone. It may also flash again when you create a new e-mail and again when you enter the recipient's address. This is normal.
- Start timing the flashing when you press the "Send" button. The flashing should last less than 20 seconds if you are not the recipient. If you sent an e-mail to yourself, the flashing could last twice as long because your phone may also retrieve the e-mail as soon as you send it.
- Make a phone call. Begin timing the moment you press "Disconnect" or "End." Again, the flashing should last less than 20 seconds.

For added assurance, conduct these tests again using a similar phone associated with the same phone carrier. The results should be comparable. If *your* phone has consistently longer transmit times, you should suspect that spyware is at work.

The SpyWarn Mobile™ timing tests alone are not conclusive proof of a spyware infection, either way. Instead, think of these findings as the twentieth warning clue.

Judgment time

To make a determination about the likelihood of your phone being bugged, you need to look at *all* the evidence you have collected. Put your intuition aside for the moment, along with the list of strange coincidences you've noticed. While these are also very important in determining whether you have a spying problem, they are not needed to make a determination about your cell phone being bugged. For now, we want to consider only *facts* that are directly related to your cell phone.

Consider whether any of the following criteria apply:

- How many of the original nineteen warning signs apply to your phone?
- Is your Duty Cycle Percentage number higher than 5 percent?
- Is your Duty Cycle Ratio lower than 20:1?
- Does your phone usually continue to transmit for more than 20 seconds after you stop using it (or at regular multiples of stopping)?

Given all the variables (different phones, features, carriers, etc.), there is no exact *yes/no* dividing line. There is no black-and-white set of statistics that prove whether or not you are experiencing a breach of privacy. Your decision, as in a court of law, will be based upon the preponderance of the evidence.

If the proof is *not* there to definitely say, "Yes, it's spyware," then bring your intuition and coincidence experiences back into the equation. Consider the advice in the "Other Phone Privacy Invasions" and "Murray's Test for Leaks™ Protocol" chapters. This will help you get to the bottom of your privacy problem even if spyware is not the cause.

If your collected evidence is telling you this *is* a spyware problem, then you are ready to move on to the next step.

YES, YOUR PHONE IS BUGGED—WHAT NOW?

You have several options. Start with the most cost-effective of these:

- Ask your service provider to reinstall your phone's software and provide a new SIM card (on GSM phones). Do not load anything onto this phone that comes from your previous phone's backup files; you may have inadvertently backed up the spyware. Instead, delete your old backup files and start fresh.
- Replace your infected phone with a new phone, complete with a new SIM card and phone number. Again, do not load anything onto this phone from your previous phone's backup files. Password-protect access to your phone. Do not let the phone out of your control.
- The 100 percent, surefire, final solution is to purchase a very basic, no-frills mobile phone. Spyware cannot be loaded onto these phones—they don't have computer capabilities or the memory capacity to hold spyware programs. Use this type of phone whenever privacy is required. Although this may not be a very convenient solution, it is a 100 percent, sure-fire, final solution.

"Should I sue the bastard?"

If you are considering legal action against your suspected eavesdropper, you need to take another step. Contact an attorney who specializes in digital technology or personal harassment.

Your phone findings, although helpful, are not enough to prove that *a particular person* was responsible for placing the spyware on your phone. You will need evidence that ties the suspect to the crime. The chapter "Murray's Test for Leaks™ Protocol" will help you develop useful circumstantial evidence to do this. Work with your attorney to develop specific tests whose results will stand up in court.

CHECKLIST: HOW TO DETER FUTURE SPYWARE PROBLEMS

Dealing with a spyware privacy invasion on your mobile communications device is a draining experience, both emotionally and time-wise. You don't want to go through this again. Following these few simple rules will help prevent future spyware attacks. It is also a good list to share with friends to help keep them safe.

www))

- **Do not jailbreak your phone's operating system software.** This is your first line of defense against spyware attacks.
- **Do not let your new phone out of your possession.** It takes a snoop only minutes to activate spyware on your phone or pull your SIM card to read the information stored on it (contacts, etc.).
- **Do not put a new SIM card into your old phone.** This will not solve the problem. Some spyware has

the capability to detect new SIM cards and will report the new phone number to the spy immediately, thus continuing your privacy problems.

- **Do not sync a new device with the old device's contacts/apps backup file.** Syncing could bring your problem back to life. You may have backed up the spyware. Delete the backup. Start fresh.
- **Use your mobile device's password feature.**
- **Set your device to lock after the shortest *time of inactivity* period.**
- **Use your SIM card's password PIN feature to prevent unauthorized access to stored information.** Here is how this security feature works: If your PIN is entered incorrectly three times, the SIM card is blocked. You can then unblock it only by entering a personal unblocking code (PUC) provided by the service operator. If the PUC is entered incorrectly ten times, your SIM card will be permanently blocked and you will have to buy a new SIM card.
- **Do not store any confidential information on your mobile device that you cannot afford to lose.** Assume there is a possibility your phone will be stolen, lost, hacked, or infected with spyware.
- **Never use any wireless device to access your bank and credit card accounts.** This includes your wireless laptop and iPad devices as well.
- **Keep current on your software updates.** They frequently include security-related improvements.
- **Download e-mail attachments only if you trust the source.** Your basic policy should be "Unknown? Leave it alone." Free ring tones, songs, and games

fall into this category. Even if your source is a trusted friend, he or she may unknowingly be passing along spyware or other forms of malware. Ask yourself, "Do I really need this?"

- **Never install pirated software on your cell phone.**
- **Monitor the Usage log built into your device.** Write down the usage at the beginning and end of the day. Keep an eye out for unexplained spikes in usage (both text and voice). This chore may be made easier with a utility usage app that logs and charts usage for you. Search your app store's Utility section using the search term "usage." Some apps will automatically notify you when a threshold limit is exceeded.
- **Turn off your mobile devices when you are not using them.** It sounds simple, but surprisingly, most people leave their devices on. If you can remove the battery, do that as well.
- **Consider purchasing a second phone that no one else knows about.** Keep it hidden, and use it only for your most important calls. Remember to turn off the caller ID function.

BONUS FOR YOU

These security tips will also help protect you from the many nonspyware, but still harmful, mobile device viruses and Trojans out there. Some are known for sending mass SMS texts and expensive MMS (multimedia) messages; manipulating your desktop; adding weird icons; or opening up your Contacts file for harvesting. All this can dramatically drain your battery, your phone bill, and your patience! Others are just malicious—they have no purpose other than to www))) disrupt your life. Called "malware," these programs attack the phone's operating system and can shut the phone down completely.

"What else do I need to know about cell phone spyware?"

To get the latest information about what spyware can do and which phones it can be used on, you need to visit the enemy. A field trip to spy shop websites is good idea—but first, read the next chapter. This particular field trip takes place in a minefield.

SPYWARE SCAMS, MISLEADING NOTIONS & "EXPERTS"

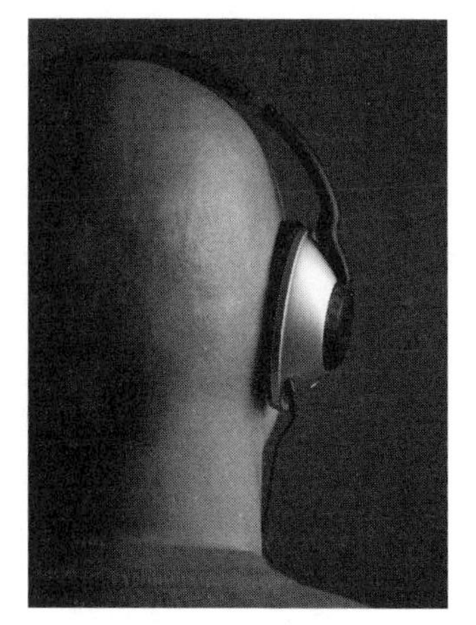

When life experiences repeat themselves, the lessons learned are often carved in linguistic stone, as an adage. There are four old adages in this chapter—warning signs along the road to personal privacy.

These days, the spotlight is on cell phone spying, and this heat attracts scam artists. Thus, our first old adage: *Sunny places attract shady characters.* You need to be careful about whom you trust when asking for advice and guidance with your privacy. Privacy is a valuable and delicate commodity. Many want to steal it. Others make money pretending to protect it. Let's look at some of the ways consumers may be getting their pockets picked.

BEWARE OF SPYWARE SCAMS

Vendors in this international bazaar come in all shapes and sizes: local spy shops, Internet "affiliate" resellers, overseas companies that actually write and sell their own software, hucksters who sell imaginary spyware. Consider another old adage: *You can't cheat an honest man*. This is very true when dealing with spyware vendors. If you do not buy spyware, your possibility of being cheated by vendors is zero.

Here is an example of what one Internet ad claims. Does this sound too good to be true? Read the hype. You decide.

> **For $99.95 and 5 minutes of your time, you too can spy on any cell phone in the world! Unleash the very latest technology that allows you to spy on any cell phone, laptop, or Bluetooth-enabled device . . . Not just one program but an entire suite of programs designed to work with old phones, new phones, smartphones, Java phones—there is a solution for any phone out there . . .**
>
> **INSTANT DOWNLOAD. PURCHASE RIGHT NOW! That means that in as little as 5 minutes, you can be checking any phone you want. Instantly you will be reading your wife's SMS/text messages, listening to your kid's phone conversation, even knowing what your boss or your neighbor is saying.**
>
> **All this is done in stealth mode. That means they will never know that you are doing it! And this**

> **works on ALL phones. Samsung, Motorola, Nokia, Ericsson, BlackBerry—absolutely all of them. It will work in all countries, on all networks. And you don't have to install any software on the phone(s) you wish to spy on! This product only needs to be installed on your own phone.**

If you're thinking this ad is making promises it can't keep, you are correct. This pitch makes no sense at all, yet hundreds of people are sending in their money and receiving nothing. What is their recourse? They have none. Who can they call for help—the police? No, that would be like complaining that your local drug dealer took your money and did not deliver the goods. What about going right to the company and demanding a refund? What company, where? All you have is an e-mail address that gets delivered somewhere on planet Earth. Good luck finding the people who cheated you.

Oh, by the way, before pressing that "Download" button, click the tiny link *waaaaaaaay* down at the bottom of their long page of spectacular promises. Here, let me do it for you.

Click.

> Legal Info: Refund Policy
> There is no guarantee that this product will do what it claims to do. The product, service, or membership referenced herein is sold with a no-refunds-allowed policy. All sales are final!

Amazing! This scam is currently being delivered via 286 websites. There is a good chance you will run across it if you look for it. Somewhere, P. T. Barnum is smiling while toasting with W. C. Fields. I can hear the glassware clinking—they would have loved this. Just think back to a few quotes they gave us that have become modern-day "adages":

> "There's a sucker born every minute."
> —P. T. Barnum

> "Never give a sucker an even break."
> —W. C. Fields

SPYWARE DETECTION SOFTWARE

Even more dangerous than bogus software is the somewhat misleading notion that you can detect and defeat all types of spyware running on your cell phone with anti-spyware software.

The public has been preconditioned by using effective computer anti-spyware to think that the same magic bullet is available for cell phones. A strong desire for this solution predisposes most people to read into advertisers' claims and to believe things that are never really claimed. Under these conditions, it is easy to be misled into making a purchase.

Unlike computer software, which is dominated by one operating system, mobile devices have several very proprietary or company-specific operating systems that do not allow examination of their code and processes. The most that software companies can do is write an application or app that may detect some very obvious changes to the original code they wrote or may identify operations that appear abnormal. Professional spyware guys know this, and they fly under that radar.

Although not an outright scam, the following is an example of a misleading ad. The Web page begins like this:

> **IS SOMEONE KEEPING SECRETS FROM YOU?**
> **Reveal All with the World's Most Powerful Spyphone!**

This is followed by ads for five different cell phone spyware packages. The sixth package, however, is called something like "Armor."

> **HIDE ALL YOUR MOBILE COMMUNICATIONS FROM SNOOPERS!**
> **Cloak all communications from specific contacts!**
> **Includes Call Manager features for Symbian 9 & Windows Mobile!**

So now you're saying, "Wow, real anti-spyware from a real, honest-to-goodness cell phone spyware company. Even comes with a 100 percent money-back guarantee!" I can hear you cooing at the thought of it.

Not so fast, bucko. Get your finger off the money button . . . now! Go read the fine print:

> **[The product] will shield all communication activity, including the ringing of an incoming SMS or phone call, from any contacts that you specify. MMS, e-mail, phone logs, and SMS from these shielded contacts are silently suppressed and stored in a hidden database for later use. [The product] also provides traditional call screening features such as Black/White lists and Allow/Disallow address book contacts.**

You may *want* to believe this spyware company is selling "a self-licking ice cream cone," as my friend John Nolan used to say in his competitive intelligence training courses (both pro- and counter-). But read this claim again. Does it *really* say it will protect you from spyware? No. Did you think it might? Yes.

Instead, this software simply shunts incoming data to a holding bin for later viewing. It works like your desk phone's "Do Not Disturb" button, which pushes your calls to voice mail. It was *not* created to thwart spyware. But, by promoting it on a spyware page and calling it by a name like "Armor," a reasonable person could be easily fooled.

Why do sellers mislead? We can only guess, but you can be sure it has to do with making money. This product, sold on its own, with a clear explanation and no reference to spyware around it, would likely not sell as well as their clever "cross-pollination" promotion scheme.

Even factoring in the 10-day, money-back guarantee, this scheme is still profitable for the vendor. Not everyone will ask for their money back. People may not even realize that the product they bought is not doing what they thought it would do. And if they do realize this, they may be disappointed, but they're still saying to themselves, "Hey, the software does *something* useful. There is no alternative product I found that will do it. And frankly, it is too much of a hassle to jump through the hoops to do anything about getting a refund."

SPYWARE "EXPERTS"

They are everywhere. From people with no credentials or formal training (like the part-time counter man at the local spy shop) to people who may have attended a one- or two-day seminar (as some private investigators do) to the true forensic specialists (mostly employed by law enforcement agencies or by private companies serving the legal profession)—everyone claims to be an expert.

So who do you need to see, and when would you need them? When your concern is, "Is my cell phone bugged?" you don't need any of these people. You have already done the right thing. You are holding the most cost-effective solution in your hands: this book.

In the previous chapter, "Spyware Detection: The Murray Method™," you conducted your own examination. There is no need to take your phone to a spy shop or a private investigator, or to send it to someone on the Internet for "an exam." Doing that will cost you much more than the price of this book, and you will learn a whole lot less. Worse, you may be given misinformation, either out of ignorance or as part of a scam to get you to spend even more money.

TIP ▪ A healthy sense of caution is better than a false sense of security.

SPYWARE AND ANTI-SPYWARE

The best way to know what you are up against is to study the enemy. Read about spyware. Get to know the latest software and features. Learn which phones each type of software can infect. Is yours on the list? If not, move on, and continue to narrow down the list of possible culprits. Learn how spyware for your phone is installed. Find out how to uninstall it. Amazingly, some spyware-mongers actually post this information publicly.

Mobile device spyware is constantly evolving and improving. The latest "news" about spyware capabilities for your particular mobile device's operating system may be found at websites like these:

- **Android:** androidspy.com
- **BlackBerry:** blackberryspyware.com
- **iPhone/iPad:** squidoo.com, iphonespy.com
- **Nokia:** spynokia.com

Be aware (and beware) that these "information" sites are usually not independent or unbiased; they appear to have strong ties to spyware sellers. They can be good for learning about spyware quickly, but understand that they are providing information from a spyware seller's perspective.

Spyware websites

- **Flexispy** (Symbian, Symbian 9, Windows Mobile 5/6, BlackBerry, iPhone): flexispy.com
- **Mobile Spy by Retina-X** (Android, BlackBerry, iPhone, iPad, Windows Mobile 5/6, Windows 2003 SE, Symbian 9): mobile-spy.com, retinaxstudios.com [9]
- **MobiStealth** (Android, BlackBerry): mobistealth.com

9 According to *The Wall Street Journal* (Justin Scheck, "Stalkers Exploit Cellphone GPS," August 3, 2010), the company's operations director says Retina-X has sold 60,000 copies of its Mobile Spy software. At $99.97 for the annual subscription, this comes to $5,998,200 per year for simply licensing a piece of software—very profitable. With this enormous income, you can understand how spyware developers can hire some of the best software designers in the business—and they do.

- **Spyera** (Nokia, Apple iPhone, Windows Mobile 5/6, BlackBerry): spyera.com

Anti-spyware software

- **F-Secure Mobile Security** for Windows Mobile and Symbian: f-secure.com
- **SMobile Security Shield** for Windows Mobile, Symbian, Android: smobilesystems.com
- **McAfee Mobile Security** (Enterprise solutions): mcafee.com
- **Symantec Mobile** for Windows Mobile and Symbian: symantec.com
- **Kaspersky Mobile Security** for Windows Mobile and Symbian: usa.kaspersky.com

Current anti-spyware software does not detect *all* types of spyware, and it needs to be on the phone *before* the spyware placement is attempted. Capabilities are updated regularly, so check the manufacturer's websites for their latest specifications and compatibility lists. Be sure to *carefully read and understand* what these products will do. From that, you will have to extrapolate what they *cannot* do.

Although you do not need any of the spyware products mentioned above, it is useful to know they exist and how they work. That way you can more easily recognize them and avoid them. The anti-spyware software is useful—if you use it *before* you have a problem.

TIP ▪ Anti-spyware products should not be relied on 100 percent.

Instructions for disabling some of these products are available in tutorials and videos freely available on the

Internet. If you use an anti-spyware product, there is the chance your spy will know how to bypass it. Anti-spyware is another brick in the security wall, but you still have to follow our checklist. So remember: Lock your phone. Lock your SIM card with a PIN code. And keep your phone in your control at all times.

TIP ▪ Beware of websites and instructional videos that are in fact spies themselves, working under the guise of fixing spyware problems.

Free advice is usually not worth the cost of admission. In this case, it is worth less than nothing. These "gurus" actually use the opportunity to plant their own spyware during the "fix-it" process.

"Why would someone want to spy on me?" you innocently ask. "They don't even know me!" The answer is, some spyware makes money by routing text messages to your phone from high-priced websites that automatically charge your phone bill. So these people are not interested in you. They are interested in your money.

Our common goal is identifying spyware and eradicating it from your cell phone. You now know that you can do this yourself. For the price of this one book, you saved yourself a trip to a sleazy spy shop and at least $100.

And did you miss anything in the process? Well, you did miss the drama of the clerk taking your phone into the back room, having a smoke, and coming back to pronounce that your phone is now "clean." You may also have missed hearing, "By the way, let me shows you dis. It ain't cheap, but youz gonna need it or da spywarez will happen again. And let me show dis tap detector for your home phone. See the lights? Push da button. Bango! If it goes green, youz clean. Red, youz dead."

There—now you didn't miss anything. Including your money.

CORDLESS PHONES AND OTHER WIRELESS DEVICES

In this chapter we leave cellular telephones behind and take a look at the other types of mobile wireless devices you may use within your home and/or workplace: cordless phones, (((www www))) wireless headsets, wireless presenter's microphones, etc.

As with cell phones, wireless transmissions from our other communications devices can be intercepted. The eavesdropper simply needs to find the signal and convert it into intelligible speech. The task is easy, when the device has an analog transmission signal. Signals from our gadgets are fairly easy for the eavesdropper to zero in on. The devices are assigned to use only a few specific frequencies, and their transmission range is usually limited to under 300 feet.

Some of the analog FM devices that are vulnerable to eavesdropping include the following:[10]

---

10 Reception distances may be even greater if the eavesdropper is using a sensitive receiver with a directional antenna.

- www))) Cordless telephones (range up to ~1/2 mile)
- www))) Wireless headsets (range up to ~300 feet)
- Baby monitors (range up to ~1/4 mile)
- Wireless video (range up to ~300 feet)
- Wireless stage/presenter's microphones (range up to ~1/4 mile)

Like cell phones, these wireless communications systems are simply radio transmitters. Your voice and/or picture are transmitted through the air on radio waves. With analog cordless phones and headsets, the threat to your privacy doubles if the person you are talking to is using a similar wireless device at his/her end of the call.

Transmissions from these systems are not unidirectional; the signal does not travel in a straight line from the transmitter to the receiver. Radio waves move like the water ripples after a stone has been dropped into a pond. A transmitter generates radio waves in all directions. These waves penetrate through walls and travel far beyond the intended receiver.

As mentioned earlier, anyone within range, using the proper receiving equipment, can listen in. The radio receivers required for intercepting these signals are readily available; so are books and websites showing transmission frequencies and tips on how to eavesdrop. This type of eavesdropping can be accomplished for less than $100 and is especially a problem in urban areas where apartment buildings house hundreds of tenants in close proximity to one another.

EQUIPMENT RECOMMENDATIONS

Whether enhancing your privacy by upgrading your communications devices, or just shopping for new gadgets, here are some recommendations to keep in mind:

- Use digital devices instead of analog ones for greatly increased—but not necessarily absolute—security.
- Buy communications products (cellular phones excepted) that transmit digitally in the 5.8 GHz or 1.92-1.93 GHz (DECT) frequency ranges. Products operating in these frequency ranges are the least prone to interception and interference from other wireless devices like Wi-Fi.
- In addition to using these frequency ranges, make sure the product employs spread spectrum or digital encryption technology.

www

GOOD COMMUNICATIONS PRACTICES

In addition to using eavesdropper-resistant technology, structure your communications to be eavesdropper-resistant as well:

- Assume someone is eavesdropping, and speak accordingly.
- Keep your calls low-profile, uninteresting, and short. Don't attract attention. Be discreet.
- Use only first names. Invent code words for sensitive topics. Avoid direct names or other identifiers.

- Remind the person you are talking with that you are using a wireless telephone, so he/she does not reveal any sensitive information inadvertently.
- Do not leave sensitive information on voice mail. Hacking voice mail is another one of the eavesdropper's tricks.
- Educate others about wireless phone eavesdropping; interception can happen on their end of your phone call as well as your end.

In addition to the noncellular communications devices mentioned above, there is one technology in particular, used in a multitude of communications products, that has especially caught the attention of hackers: Bluetooth®.

BLUETOOTH® EAVESDROPPING

Bluetooth® is a very helpful wireless digital technology. It is used in headsets, keyboards, mice, speakerphones, and many other devices. The common transmitting range is approximately 30 feet (10 meters), although some iterations of Bluetooth transmit up to 300 feet. The *actual* range of interception of any Bluetooth transmission, however, can be extended by snoops who use high-quality receivers and directional amplified antennas.

According to Bluetooth.com, "Bluetooth technology operates in the unlicensed industrial, scientific, and medical (ISM) band at 2.4 to 2.485 GHz, using a spread spectrum, frequency-hopping, full-duplex signal at a nominal rate of 1600 hops/sec." This *sounds* very secure.

However . . .

Bluetooth devices are targets for hackers and eavesdroppers. Eavesdropping on people with Bluetooth headsets and hands-free car systems appears surprisingly easy to accomplish according to the hackers who demonstrate their attacks on YouTube.com and other websites.

There are several well-publicized Bluetooth attacks that can cause privacy problems. These include Bluejacking, BlueSnarfing, BlueBugging, HelloMoto, BlueSmack, and BlueSniping. Each attack does something slightly different from the other, but generally speaking, you can expect eavesdropping, theft, of information stored on your phone, and use of your phone when you are not using it, for reasons like making fake calls to 911. (And you really do not want "Bluetooth" to mean the guys in blue—the cops—gnashing their teeth at you!)

The National Security Agency (NSA) recognizes the security risks too. This is what the agency has to say:

> Bluetooth is a short-range, low-power wireless technology commonly integrated into portable computing and communication devices and peripherals. Like any wireless technology, Bluetooth introduces a number of potentially serious security vulnerabilities. These vulnerabilities may lead to the compromise of the device and those networks to which it connects. Proper use of standard Bluetooth security features, however, should provide adequate security for many unclassified applications.

BLUETOOTH SECURITY MECHANISMS

Bluetooth links use optional pre-shared key authentication and encryption algorithms that are widely considered acceptably strong when both implemented and used correctly. The strength of Bluetooth security relies primarily on the length and randomness of the passkey used for Bluetooth pairing, during which devices mutually authenticate each other for the first time and set up a link key for later authentication and encryption. Also important for overall Bluetooth security are discoverability and connectability settings. These settings control whether remote Bluetooth devices are able to find and connect to a local Bluetooth device. Optional user authorization for incoming connection requests provides additional security.

BLUETOOTH VULNERABILITIES

Bluetooth is, by design, a peer-to-peer network technology, and [it] typically lacks centralized administration and security enforcement infrastructure. The Bluetooth specification is very complex and includes support for over two dozen diverse voice and data "profiles" or services. Some of these include headset, cordless telephony, file transfer, dial-up networking, printing, and serial port profiles. In addition, designers have implemented Bluetooth using a wide variety of chipsets, devices, and operating systems.

This variety in foundational elements results in user interfaces, security programming interfaces, and default settings that also vary widely. Because of these complexities, Bluetooth is particularly susceptible to a diverse set of security vulnerabilities. Publicly documented Bluetooth attacks involve identity detection, location tracking, denial of service, unintended control and access of data and voice channels, and unauthorized device control and data access. As an example, researchers have shown that Bluetooth headset use can compromise devices in multiple ways. This compromise is due to the headset profiles' support for powerful telephony signaling commands and the all-too-common use of weak fixed passkeys (typically "0000").

The website http://www.bluetooth.com/English/technology/works/security/pages/protecting.aspx contains a general discussion of Bluetooth security and vulnerabilities, and http://www.trifinite.org contains detailed descriptions of Bluetooth attacks along with downloadable audit and demonstration software. While the proper use of existing Bluetooth security mechanisms can reduce Bluetooth security risks to a level acceptable for unclassified use, Systems Network Attack Center (SNAC) research will continue into Bluetooth vulnerabilities and secure Bluetooth design for use in sensitive and classified security domains.

BLUETOOTH USE IN THE DEPARTMENT OF DEFENSE

As a result of comprehensive security evaluations conducted by SNAC and in close partnership with vendors, the Defense Information Systems Agency (DISA) has begun approving the unclassified use of Bluetooth Common Access Card (CAC) readers with BlackBerry, Windows Mobile, and Windows XP Service Pack 2 devices. The SNAC develops stringent security requirements for DoD Bluetooth use and verifies the secure implementation of each approved solution. DISA then publishes security requirements and recommended configuration guidance in matrices and checklists at iase.disa.mil/stigs/checklist. Standard commercial Bluetooth headsets remain prohibited in the DoD for the reasons listed above.

BLUETOOTH SECURITY RECOMMENDATIONS & PRECAUTIONS

Both users and Bluetooth application developers have responsibilities and opportunities to minimize the risk of compromise via Bluetooth. Users should follow these best-practice security guidelines:

- Never use standard commercial Bluetooth headsets. (Secure Bluetooth headsets are available.)
- Enable Bluetooth functionality only when necessary.

- Require and use only devices with low-power Class 2 or 3 Bluetooth transceivers.
- Keep devices as close together as possible when Bluetooth links are active.
- Independently monitor devices and links for unauthorized Bluetooth activity.
- Make devices discoverable (visible to other Bluetooth devices) only if/when absolutely necessary.
- Make devices connectible (capable of accepting and completing incoming connection requests) only if/when absolutely necessary and only until the required connection is established.
- Pair Bluetooth devices in a secure area using long, randomly generated passkeys. Never enter passkeys when unexpectedly prompted for them.
- Maintain physical control of devices at all times. Remove lost or stolen devices from paired device lists.
- Use device firewalls, regularly patch Bluetooth devices, and keep device anti-virus software up to date.
- Comply with all applicable directives, policies, regulations, and guidance.
- Subject Bluetooth solutions and deployments to independent security audits by qualified evaluators.

The following design advice that the NSA gives Bluetooth application developers is even more revealing about the agency's security concerns:

- Eliminate or disable support for the Headset and Hands-Free profiles unless such links are adequately secured using the techniques described.
- Passkeys should be at least eight digits long. Passkeys must not be valid indefinitely.
- Use configuration and link activity indicators like LEDs or desktop icons.
- Use nondescriptive Bluetooth device names on each device, and identify all paired and connected Bluetooth devices by hardware (MAC) address.
- Require user authorization for all incoming connection requests, and don't accept connections, files, or other objects from unknown, untrusted sources.
- Program each device to initiate Bluetooth authentication immediately after the initial establishment of the Bluetooth connection (also known as Security Mode 3, Link Level security).
- Program each device to initiate 128-bit Bluetooth encryption immediately after mutual authentication. Layer FIPS-certified cryptography (Federal Information Processing Standard) atop Bluetooth cryptography for defense in depth.
- Store link keys securely, and regularly keep them under encryption.
- Remove the user's ability to control Bluetooth settings that could possibly circumvent security features.

- Enable each Bluetooth service only when needed. Permanently remove, or disable, all unnecessary Bluetooth services.
- Digitally sign all Bluetooth firmware, driver, and application software. Verify that no unauthorized software applications use Bluetooth application programming interfaces.
- Prohibit the user from changing Bluetooth security features.

Improvements in the standard and in how manufacturers now implement it have reduced eavesdropping vulnerabilities. But . . . as the hacking programs and techniques evolve, expect more attacks.

All the preceding information can be boiled down to two security tips for Bluetooth users:

Bluetooth Tip No. 1: Turn off Bluetooth whenever you don't really need it.

Bluetooth Tip No. 2: Customize your Bluetooth security PIN—the longer, the better.

You now know the most common privacy and eavesdropping vulnerabilities of your communications tools. Next, we take a look at an equally important, but often overlooked, aspect of combating snoops and spies: paying attention to the secondary information security loopholes.

OTHER PHONE PRIVACY INVASIONS

Now you that have switched your analog cordless phones and headsets to digital, avoided spyware installation on your phone, and run the Duty Cycle test—are you protected? Yes and no. These were very important first steps. You have dramatically reduced the possibility of eavesdropping by interception of your wireless signal, but there is more to consider.

Not all eavesdropping depends on intercepting the radio signal or installing spyware on your phone. It is possible that what you say is being intercepted elsewhere along the route of your call. If you still suspect that your calls are being intercepted or your phone is wiretapped, consider the path your voice takes during a phone call. On your end of the call:

- Is anyone around who can simply overhear what you say?

- Is it possible that the room has a listening device in it?
- Could your home or building phone wiring be tapped?

Cordless phones are *only* wireless between the base and the handset. The base station plugs into wiring that eventually leads back to the phone company. Once your call reaches the phone company, it is available to some of their employees and to law enforcement, under very strict guidelines. It is extremely unlikely that the average, law-abiding person would be eavesdropped on at the telephone company.

Meanwhile, at the other end of the call:

- Will the person you are talking to keep the details of your conversation confidential?
- Is he or she using any type of analog phone? (If so, both sides of the call are vulnerable to interception.)
- Is it possible there is a wiretap on his/her end of the line?
- Is it possible the area he/she is in is bugged?
- Did someone else near him/her overhear the conversation?

Here are some tips to help you avoid privacy invasion when using any sort of wireless device:

- Keep all the above possibilities in mind.
- Avoid using an analog wireless device whenever possible.

- When having conversations, try to be discreet and uninteresting.
- Avoid using specific names, dates, and places unnecessarily.
- Test for leaks. Occasionally release a bit of information that, if overheard, would cause an eavesdropper to do something to reveal him- or herself.
- If your calls are extremely sensitive, consider adding encryption to your phone for end-to-end privacy.

As with every rule, however, there is an exception: If your calls involve business or governmental information, the rules change dramatically and you will need to engage the services of a counterespionage specialist.

For everyone else, following the steps outlined in the next chapter will do just fine.

MURRAY'S TEST FOR LEAKS™ PROTOCOL

Contrary to what people commonly think, having spyware on a cell phone is not their real problem. It is just one symptom of a much darker disease—privacy invasion. Your fight is not against spyware; it is against the person who put it there. The solution is not only to prove that your phone is infected with spyware, but to prove who is illegally using it to invade your privacy. Accomplish this, and chances are good that your problem will go away.

Understandably, people in general have a knee-jerk reaction to being eavesdropped on: Find the spyware, bug, wiretap, or whatever device is being used. Buy a counterspy gadget. Call in an expert to search and destroy.

Stop! Take a deep breath. Carefully read this chapter. For most people, the best first step toward solving their privacy invasion problem *does not* require hiring a technical specialist to search for spyware on their cell phone or conduct a sweep for bugs. *You* can nail your spy yourself, without spending a fortune. And you stand a very good chance of solving your privacy problem once and for all.

Murray's Test for Leaks™ is a protocol created by Murray Associates to solve espionage and information loss problems faced by its corporate and government clients. The success rate for solving personal privacy issues using this method is high. Follow the directions as outlined, and you very likely will get to the root of your problem.

Note: If you feel your life is in danger, or a crime is being committed, contact law enforcement for help—RIGHT NOW. Otherwise, you should be able to accomplish your goal of restoring your privacy using your own abilities, technology, intuition, and common sense.

A common question is, "Why not hire a professional first?" There are three *really* good reasons:

1. Even if you find spyware on your cell phone, all you actually have is a piece of hardware with altered software . . . and an eavesdropper who may now know you are suspicious, who is free to do it again, and who will do a better job of hiding his/her snooping the next time.
2. Your information loss *may not even involve electronic surveillance*. Leaks occur in other ways, too. It makes good sense to get these possibilities out of the way first.

3. If the "detective" you hire is not technically competent enough find the spyware, all you really have is continued uncertainty and a lot less money. Besides, *you* can make an educated determination simply by using this book. Your level of cost-effectiveness will exceed that of the "experts" out there.

TIP ▪ If you really do need additional assistance, favor the people who offer themselves as "specialists" over those who proclaim "expert" status and/or have only "law enforcement" experience.

THE FOUR STEPS

The four steps required to solve an information loss problem are as follows:

1. Demonstrate to others that your information losses are specific and deliberate privacy invasions, not just coincidences.
2. Prove that your information was obtained by a specific person via illegal means.
3. Determine what methods and/or technical techniques were used to illegally abridge your privacy or obtain your information.
4. Show a *cause-and-effect relationship* between the criminal and the crime. This is called *circumstantial evidence*—evidence that indirectly proves a fact.

www))

For example, you call a friend and tell him you will meet him at the mall in a particular store. There is something specific you need to pick up. You only mention this once during the phone call and never mention it again. Your friend, of course, is in on the test and also knows not to mention it either. Follow through and go to the mall, but don't go anywhere near the store you mentioned. Doing this will help you determine if your spy was using visual or electronic surveillance techniques. If word of your supposed trip gets back to you, listen carefully. Did this person mention that you went to the store? If so, suspect electronic eavesdropping. Did he or she mention only that you went to the mall? Perhaps visual surveillance or a vehicle tracking device provided this information. In either case, you *know* that the person who mentions it can bring you one step closer to your spy. He or she may have talked to the spy, or may even *be* the spy. Don't let on that you suspect him/her yet; there is more to do before springing the trap.

"How do I accomplish these steps?"

Suspected information leaks can be verified and tracked to their source. Conduct each one of the following tests using a different and very specific piece of information each time. *It is imperative that this information is* absolutely *associated with only one specific test.* There should be *no* other excuse your snoop can use to explain how he/she knew. In addition to proving a case against your snoop, his/her *method* of spying will be revealed by virtue of which tidbit of leaked information surfaces.

It can be challenging to make up and distribute this "bait" information. Believability is a key element. It also must be interesting enough to cause a documentable reaction—your rat has to either say or do something that could *only* have been prompted by the bait information. Plus, it must not smell to your rat like bait. And you will need a lot of it.

There are several tests to be done. You need at least three positive reactions to show the person's ongoing *intent* to commit the crime of eavesdropping—intent that cannot be explained away as "a coincidence."

Run your tests in each of the following scenarios:

1. **Loose lips and false friends.** Although we all hate to think that our confidences have been betrayed by trusted friends and loved ones, this is the most common way information leaks out. Selectively salt your one-on-one conversations with false or unique bits of information—one tidbit per person being tested and a different tidbit for each person. This approach reveals loose lips and false friends if the information gets out.
2. **Wiretaps.** Prearrange a telephone conversation with a trusted person who is using a phone that would *not* be targeted for wiretapping. *Each of you* should mention one interesting and absolutely unique item. If word gets out, you can suspect spyware (if on a cell phone) or a wiretap. Conduct this test on each phone you normally use, using a different piece of information each time. If your piece of information is the *only* one that gets out, it then becomes possible that a room bug picked up your side of the conversation. If this is the case, test no. 3 becomes very important.
3. **Room bugs.** Pretend you are talking on the phone (do not really use the phone at all), or invite a trusted

friend over to have a face-to-face discussion. Discuss something that is likely to get back to you. Do this in all the rooms you use, using a different piece of information in each room. Because bugs can hear as well as humans, a bug in a central location will hear conversations in nearby areas, too. Finding the exact room the bug is in may require some winnowing. Positive results from this test are an indicator of room bugs or hidden voice recorders. If you get a very quick "positive" result from this test, logically a room bug becomes a more likely suspect than a voice recorder.

4. **Intruders.** You think no one else has a key to your home or office, yet you have the feeling your papers, letters, and files are being read. You could set up a hidden camera to see who comes and goes, but try this simple test first: Write an *interesting* note to leave on your table or desk overnight. Use a different note every day. Position it so that it has to be moved to be fully read. Upon your return you will notice if it has been moved. This proves you were visited by a snoop, and on which night. If word gets back to you about the note, you will also be on the way to discovering the intruder's identity.
5. **Garbage archeology.** Write an interesting note each night to leave crumpled or torn in half in the trash can. This detects surveillance via garbage archeology, a very common spy trick.
6. **Leakers.** Create several copies of an interesting letter or business document. In each one, slightly alter a

bit of information or punctuation. Distribute them to your suspects (remembering, of course, who has which copy). The copy that surfaces identifies your traitor.

THE END RESULT

Congratulations! You will now know *for certain* whether spying is taking place and what method is being used against you. Most important, you will have gathered valuable evidence to that effect. This is essential for legal proceedings against your snoop, which should stop him/her once and for all.

Getting positive results from your tests will provide several specific outcomes:

- Proof that a concerted effort is being made to invade your privacy, thus eliminating coincidence as the cause of your concerns
- A strong indication as to whether or not the loss was caused by cell phone spyware or another form of electronic eavesdropping
- A strong indication as to which method of eavesdropping or spying is being used against you
- The general location of the spyware, bug, wiretap, or voice recorder (assuming your test results point to electronic surveillance)
- And, most important, circumstantial evidence that ties your suspect to the crime—*essential* for eliminating the problem once and for all

The final step

If you fear for your safety or that of anyone else, or if a crime has been committed, contact law enforcement for help. Otherwise, take your documentation to an attorney who specializes in digital technology or personal harassment, and get a recommendation on how to proceed with your case (within the criminal judicial system, the civil court system, or both). Your attorney may recommend an immediate, professional technical surveillance countermeasures (TSCM) inspection—a bug sweep—to document and recover the physical evidence of the electronic surveillance against you. If so, rather than doing it yourself, have the attorney contact a specialist for further details on how to proceed.

CHOOSING A SECURE CELL PHONE

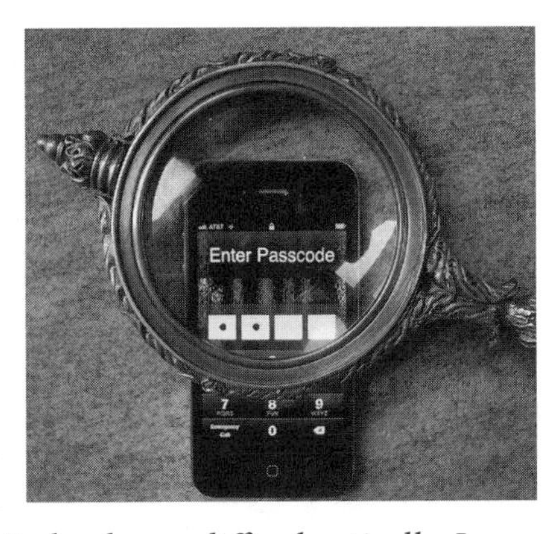

You are doing your part to protect your privacy. But what about your cell phone? Can you trust it?

You have to face the possibility that your phone may be promiscuous. Like mates, not all cell phones are created equal in terms of moral fiber. Their internal security levels can differ drastically. In short, the make and model of the cell phone you use affect the level of your personal privacy.

Whether you are considering a new phone or just giving your current phone a security evaluation, the following checklist will help you select the right mate.

CHECKLIST: BASIC SECURITY CONSIDERATIONS

When choosing a secure cell phone, be sure yours offers at least the following basic features:

- A base operating system with a proven track record for tight security
- Prevention against unauthorized access to and/or use of your phone
- Protection of transmitted data using network security and encryption
- Protection of stored data using encryption and remote destruction

How you use your phone will determine the depth of security you will require—how many of additional security items are important to you. If you're the average person, and the most secret data stored on your phone is within a grocery list app, quality password protection is the item to focus on. If you're a business executive with competitors and/or foreign government agencies breathing down your neck, you want it all—including the *Mission: Impossible* smoke coming out of the earpiece when you have been compromised. The middle-ground solution for most people involves the following:

- Using passwords for access to the phone and its SIM card
- Restricting access to the phone by others
- Never downloading suspicious software
- Knowing how to remotely erase all stored

information should the phone be lost or stolen (if the phone supports this capability)

Keep in mind that not every phone can support the full range of security options. You may have to do some further research to learn about some of the more esoteric items and their current levels of effectiveness. Your likely goal, however, is to have as many security features crammed into your phone as possible.

CHECKLIST: SECURITY QUESTIONS

Following is a checklist about specific security options that are available in today's market. These questions are definitely worth asking when you are in the process of choosing and purchasing a new cell phone:

- Does this phone have strong password protection? Some phones allow custom passwords that are longer than the standard four-digit password.
- Do the passwords expire? Change is good.
- Can passwords be reused? Not so good.
- Will the phone lock up after a maximum number of failed password attempts?
- Will it automatically require a password after a defined period of dormant time?
- Will the phone automatically erase all data if the wrong password is entered too many times?
- Can you remotely erase your data if the phone is lost or stolen?
- Can you erase your data easily in case you want to sell your phone?

- Is the phone's data stored and transmitted in an encrypted state?
- Does the phone handle the common network security, including RSASecurID, Certificate-based authentication, CRYPTOCard, WPA/WPA2 Enterprise when using Wi-Fi (802.1X), Cisco IPSec, SSL/TLS with X.509 certificates, L2TP, and PPTP VPN protocols?
- Does the phone handle platform security measures, such as common crypto APIs, mandatory code signing, keychain services, and runtime protection?
- Does the phone support Microsoft Exchange ActiveSync security policies?
- Does the phone support Microsoft Exchange Server 2007 security policies?
- How are security policies enforced: over the air? Directly, with supervision? Can they be overridden? If so, is all the data automatically erased?
- How are apps handled? Can just any app be added, or are they pre-screened and approved by the phone's manufacturer or the operating system's developer?
- Can app usage be restricted if necessary?

GOING ONE STEP BEYOND BASIC

Most people require some combination of these basic security features. Meanwhile, others require a step up in protection, which means the full spectrum of safety functions: end-to-end encryption. This is what governments and some businesses use when privacy is an absolute necessity and money is no object.

Cryptophones have been certified by the National Security Agency (NSA) to protect classified information to the level of "top secret." The cost per phone is in the $1,000 to $2,000 range. Remember, to hold an encrypted conversation, *both* parties to the call must use compatible encrypted phones.

The trade-offs for this level of protection are usually a lack of "style," a loss of features, and (of course) a lot of money. To see some of these bulletproof bricks, conduct an Internet search using their names: Sectera, CryptoPhone, TechLab 2000, TopSec GSM, and TopSec Mobile.

Add-on software security packages are also available for most phones. The prices for these packages are lower, and installation is fairly easy. Again, to use encryption, both phones require the same software. Some currently popular brands of add-on encryption software include PrivateWave Italia, SecureGSM, Sigillu, PhoneCrypt, Android Privacy Guard, and Cellcrypt.

Once your research is complete, you will own a clean cell phone with good security features that meet your needs. All is well . . . except when you need to retrieve some information off your old phone. For this, you will require a little do-it-yourself work, as described in the next chapter.

DO-IT-YOURSELF FORENSICS

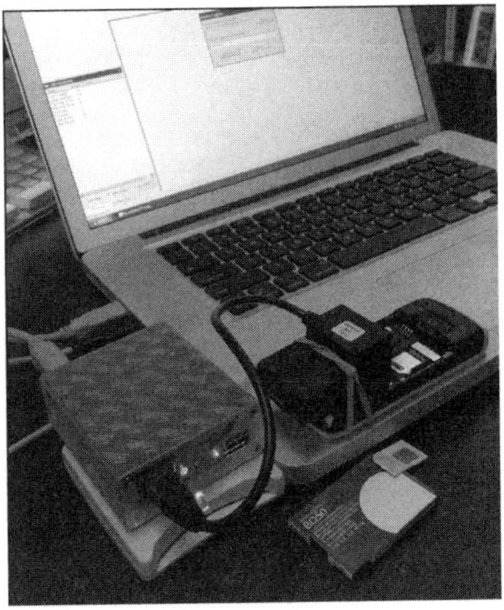

If you have a cell phone in your possession and want to retrieve data from it for your own legitimate personal needs, there are a few *CSI*-type things you can do:

- **Go through every menu item.** Take photos or make notes of what you see on the screen.
- **Use the syncing program** provided by your phone's manufacturer.
- www)) **Buy a SIM card reader** and download the information stored on the SIM card. Other attached media,

such as a MicroSD card, may also have information you will want to download.
- **Purchase data recovery software for your phone.**
- **Contact a forensic examiner** who specializes in mobile technologies for additional assistance.

One easy first step you can take is to capture—by photographing or taking notes—the visible information available on all the menu screens. Although time-consuming, this does not require any special equipment or skills. You can also use the phone's own syncing program. Torturing secrets from your little clam this way is as simple as hammering it with a few keystrokes. Best of all, this approach is cost-effective. Many manufacturers—Apple iOS, Nokia Symbian OS, RIM BlackBerry OS, Microsoft Windows Phone OS, Linux, Palm WebOS, Samsung Bada, Nokia Maemo, Google Sync—allow you to download their syncing software from their websites for free. ((www

Using a SIM card reader and a MicroSD card reader (if your phone uses this type of media) is a good next step and is not hard to do. Readers range in price from $5 to $200; the better ones come with detailed instructions. If you think this is beyond your skill level, just ask someone who is a little more computer savvy; he or she should be able to help you. You do not need to hire a specialist for this unless you think you will be taking your case to court. If so, work with an attorney and an expert witness of his/her choosing. ((www

Here is a list of what you can find on a SIM card:

- **The international mobile subscriber identity (IMSI) number.** It uniquely identifies the card on the cellular network. The number is made up of three parts: a 3-digit mobile country code, a 2-digit mobile network code, and a (up to) 10-digit mobile subscriber identity number.

- **The mobile subscriber ISDN number (MSISDN).** The mobile telephone number associated with the SIM card. (ISDN stands for "Integrated Services Digital Network.")
- **Any SMS or text messages** that were received and saved to the SIM card. Outgoing (sent) messages are *not* saved to the SIM card. (Recently erased text messages may also be retrieved, with the help of a forensic specialist.)
- **The Contacts directory.** Contact information (name and number) may be stored on the SIM card or in the phone's memory.
- **A SIM personal identification number (PIN).** This code protects the SIM card from being used by another person or in another phone if it is lost or stolen.

The next step up the difficulty ladder is cell phone data recovery software. An Internet search will locate the latest packages available for your phone. Some software even comes pre-loaded on USB thumb drives, which are branded as "recovery sticks." The cost of simple software programs ranges from $50 to $200; the full professional forensic kits are in the $2,000-to-$4,000 range. While this is above the average person's skill level, anything can be learned if you have the desire.

Again, should you need your phone's internal information for court evidence purposes, contact an attorney who specializes in digital technology. Ask him or her to locate and vet a reliable cell phone forensic examiner for you. It is likely that your attorney already knows the right person, from work on past cases. Seeking good advice based on "certifications" alone can be risky. Cell phone forensics is a new science. It will be a while before a universally respected and accepted certification process emerges.

If the analysis of cell phones appeals to you and you would like to learn more (without paying more), download *Guidelines on Cell Phone Forensics: Recommendations of the National Institute of Standards and Technology* (NIST Special Publication 800-101). It is a 108-page report (found at http://csrc.nist.gov/publications/nistpubs/800-101/SP800-101.pdf) filled with everything you need to know to get started. The Executive Summary presents the current state of cell phone forensics as follows:

((www

> Mobile phone forensics is the science of recovering digital evidence from a mobile phone under forensically sound conditions using accepted methods. Mobile phones, especially those with advanced capabilities, are a relatively recent phenomenon, not usually covered in classical computer forensics. This guide attempts to bridge that gap by providing an in-depth look into mobile phones and explaining the technologies involved and their relationship to forensic procedures. It covers phones with features beyond simple voice communication and text messaging and their technical and operating characteristics. This guide also discusses procedures for the

preservation, acquisition, examination, analysis, and reporting of digital information present on cell phones, as well as available forensic software tools that support those activities.

TIP ■ Prevent someone from CSI-ing your SIM and/or microSD card by placing security tape over it. Security tape has an adhesive backing (or frangible body) which can only be applied once. If removed, the tape will either break or show tampering. Serialized seals, like the ones shown, are available at www.spybusters.com. You can also make seals yourself using a Brother P-Touch machine and their security tape cartridge (Brother TZSE4).

Any discussion of electronic surveillance is not complete until the applicable laws are discussed. In the next chapter, we tackle some of the more common questions and list the applicable laws in case you want to research them further.

Security seals protect SIM and memory cards from tampering.

LEGAL ISSUES

Eavesdropping sparks many legal questions of all levels of complexity, starting with, "Is it legal?" (No, not in the United States and most other countries.) Some of the more frequently asked questions are listed in this chapter; the answers are what you likely would hear from knowledgeable security consultants in the United States.[11]

PRIVACY LAW

Over the years, the following questions about eavesdropping and wiretapping burp up over and over again. Answering them is not easy. They do not have simple yes-or-no answers. Generally speaking, the laws aimed at government's behavior toward privacy and eavesdropping are more strict than those regulating the behavior of private individuals. It is the latter, however, that concern us.

11 The information within this chapter is not legal advice. Before making any important decisions involving anything you read here, you need to consult with an attorney who specializes in cases involving privacy law.

Privacy law, especially where technology is involved, is still evolving. Along the way, interpretations of existing laws are being determined in the courts. Their decisions vary from district to district and from state to state. Often these decisions are exact opposites of each other.

Giving advice about cell phone privacy or predicting a court decision is like stepping into a minefield. While these questions require professional legal advice, the comments provided here may become outdated with time as new laws emerge. Speak with a knowledgeable attorney for the most accurate, up-to-date advice.

"Can the person I am talking to legally record my conversation without my knowledge?"

Want the quick answer? Well, there is no quick or obvious answer.

United States federal law says *yes*, as long as at least one party to the conversation agrees to the recording. Obviously, that would be the person doing the recording.

Some states follow the federal government's one-party rule, while other state laws are more restrictive, specifying that *every* person in the conversation must consent to the recording being made. There are no state laws, however, that are less restrictive than the federal law. The rule is that either one party or all parties must consent; when *no parties* consent (without a court order), that always constitutes illegal eavesdropping.

Further complications occur when the persons talking are in different states. The courts have decided this type of case both ways over the years. In some cases, federal law has taken precedence. In other cases, it depended on the law of the state in which the recording was done or in which the phone call originated. Confused? Join the club.

"I'm the boss (homeowner, landlord, etc.), and I own the phone. So I can put spyware on the cell phones I give my people and wiretap the phones on my property, right?"

Quick answer: Don't do it.

The federal law allowing employers to monitor employee communication "in the ordinary course of business" provides a very narrow exception. Always remember, state laws could be more restrictive. Do not do anything like this without obtaining professional legal advice first (and second, setting aside bail and defense fund monies).

"I can bug my spouse's phone because of spousal immunity law, right?"

Quick answer: Don't do it.

There is no spousal immunity law, and "spousal immunity" doesn't even mean what you may think it means. It is simply one of the Federal Rules of Evidence, and it says:

> In a criminal case, the prosecution cannot compel the defendant's spouse to testify against him. This privilege only applies if the defendant and the spouse witness are currently married at the time of the prosecution. Additionally, this privilege may be waived by the witness spouse if he or she would like to testify.[12]

12 Cornell University Law School, http://topics.law.cornell.edu/wex/spousal_immunity.

What you may be thinking about are the rulings by the United States Court of Appeals (Second and Fifth Circuits). It is their opinion that the Title III electronic surveillance laws *do not* apply to interspousal wiretapping. Conversely, the Fourth, Sixth, Eighth, Tenth, and Eleventh Circuits have held that Title III *does* apply to interspousal wiretapping. Your case could be decided differently.

"I certainly have a right to monitor my family's communications, right?"

Quick answer: Don't do it.

The court decision that is often cited as allowing an intrafamily exemption to the eavesdropping laws, *Simpson v. Simpson*, has been largely rejected by subsequent rulings. On the other hand, many courts have accepted the theory of "parental vicarious consent" when deciding cases about parents monitoring their children's conversations. Before crossing this line and placing spyware on any family member's phone, however, you should consult with an attorney. Seriously consider the many other ramifications of this action, not least of which is, what if they find out?

"He/she put spyware on my phone, so I can do it to him/her. Fair is fair, right?"

Quick answer: Wrong. This is getting silly now. Two wrongs do not make a right, right?

The focus of anti-spyware law is generally protection of the unwitting surfer from the commercial marketing sharks. Their spyware secretly collects consumer information that helps them circle in on their sales prospects.

Secondarily, the laws consider data privacy: theft via data fraud, malicious harm to data, and spam. Little attention though has been given to our concerns—communications privacy.

Congress and the Senate began working on passage of a federal anti-spyware law as far back as 2003. In its last incarnation it was called the "Enhanced Consumer Protection Against Spyware Act of 2005." One would think everyone would be in favor of a law like this, but that is not the case. Heavy lobbying by online marketing firms wanting exceptions for themselves, *and* the preemption of state laws, doomed its passage. As of this writing, no federal U.S. law prohibiting spyware of any kind exists.

Many states, however, do have anti-spam/anti-spyware laws. Most include wording like: "To combat spyware, malware, and other malicious software, the Act prohibits the installation of computer programs without the consent of the computer's user or owner." Fortunately, this wording can also be interpreted to include spyware meant for personal surveillance and eavesdropping. To be more effective, future laws will also need to borrow a phrase from electronic surveillance law:

> Manufactures, assembles, possesses, or sells any electronic, mechanical, or other device, knowing or having reason to know that the design of such device renders it primarily useful for the purpose of the surreptitious interception of wire, oral, or electronic communications, and that such device or any component thereof has been or will be sent

> through the mail or transported in interstate or foreign commerce; or places in any newspaper, magazine, handbill, or other publication or disseminates by electronic means any advertisement.
> (US Code, Title 18, Part I, Chapter 119, § 2512)

In December 2010, Canadians passed a federal anti-spyware act with an incredibly long name: "An Act to promote the efficiency and adaptability of the Canadian economy by regulating certain activities that discourage reliance on electronic means of carrying out commercial activities, and to amend the Canadian Radio-television and Telecommunications Commission Act, the Competition Act, the Personal Information Protection and Electronic Documents Act, and the Telecommunications Act." This shows that uniform legislation is possible. It also gives hope that as the cell phone spyware problem grows, doing something about it will become politically valuable to politicians—especially when a scandal erupts due to spyware on one of their phones.

ADDITIONAL LEGAL INFORMATION[13]

((www

1978 Foreign Intelligence Surveillance Act ("FISA," 50 U.S.C 1801 et seq) ((www

A Practical Guide to Taping Phone Calls and In-Person Conversations in the 50 States and D.C. ((www

Communications Assistance for Law Enforcement Act ("CALEA", Public Law 103–414, 47 U.S.C. 1001–1010)

13 Links available at www.spybusters.com/Cell911.html.

www))) The Electronic Communications Privacy Act of 1986 (“ECPA,” 18 U.S.C. 2701)

www))) The Wire and Electronic Communications Interception and Interception of Oral Communications Act (formally known as the “Title III” Wiretap Act, 18 U.S.C. §§ 2510–2520)

THE TRUTH ABOUT CALLER ID

Caller ID can increase your telephone privacy, as you will see in the next chapter, But first, you need to learn a little more about it. What you learn may surprise you, and it may disappoint you. Ultimately, though, it will help you.

MYTH VS. REALITY

Let's address the myths about caller ID—and present the reality.

"Does caller ID even show the real number?"

Myth. The caller ID phone number that you see on your phone is the caller's real phone number.

Reality. It might be the caller's real number. Then again, it could be a lie. Here is what the FCC (Federal Communications Commission) has to say about it (http://www.fcc.gov/cib/consumerfacts/callerid.html):

Caller identification, or "caller ID," allows you to identify a caller before you answer your telephone. It is an optional telephone service, available from your local telephone service provider for an additional monthly fee. A caller's number and/or name are displayed either on your phone (if your phone has this feature) or on an external display unit that you can buy separately. The number and/or name will appear on the display unit or on your phone after the first ring. This service also lets you identify yourself to the person you are calling.

Caller ID service, however, is susceptible to fraud. Using a practice known as "caller ID spoofing," disreputable parties can deliberately falsify the telephone number relayed as the caller ID number to disguise the identity and originator of the call.

By spoofing caller ID, the caller circumvents a number of security and privacy safeguards, such as *57 (call trace), *69 (last call return), anonymous call rejection, and detailed billing. Spoofing may also be used to break into voice mail systems that rely on caller ID for security. Spoofing is not just a numbers game either. In addition to sending fraudulent caller ID numbers, phrases like "New Jersey Call," "AIDS Clinic," and worse may also be transmitted as fake identifiers.

Spoofing is usually accomplished by programming the spoof into ISDN phone system settings (mostly used by businesses) and into open-source Voice over Internet Protocol (VoIP) telephone software. It can also be accomplished easily by using a spoofing service provider.

Spoofem.com, for example, provides this type of service:

> Call any number, and have any number show up in the other person's caller ID. You can change your voice to sound like a male or [a] female. You can even record a call and, when you hang up, it will e-mail the call to you! All calls are untraceable to the person you are calling.

Spoofem.com also provides a similar service for spoofing e-mails and text messages. Oh, by the way, they do appear to have a bit of a conscience—at least they post this information on the website:

> Although Spoofem.com offers a service that can help protect a caller's identity, it can also be used as a tool for phone abuse. Knowing this, we have decided to provide a service to allow you to block your number from being spoofed by phone abusers. When you submit your information, not only will your numbers be blocked by Spoofem.com, but your information will be forwarded to other telephone spoofing providers.

Right . . . for a fee. Another self-licking ice cream cone.

Spoofcard.com's service is similar to Spoofem.com's, but it only offers the basic spoof, voice change, and call recording features. Here is how it works:

> SpoofCard offers the ability to change what someone sees on their caller ID display when they receive a phone call. Simply dial SpoofCard's toll-free number

> or local access number in your country and then enter your PIN. You'll then be prompted to enter the destination number followed by the phone number to appear on caller ID. It's that easy!

SpoofCard also offer a dashboard widget for the Mac OS and apps for most cell phone operating systems. International spoofing is also offered.

Telespoof.com is yet another provider of basic spoofing, voice change, and call recording services.

Spooftel.com's pitch is simple:

> SpoofTel offers you the ability to change or "spoof" your caller ID. This is the phone number you see when you receive an incoming call to your cellular or landline telephone. With our caller ID spoofing service, you can show any phone number you wish on [the recipient's] call display. You can change your voice to male/female, record the conversation, (and) spoof SMS text messages.

Spooftel also offers a free desktop spoofing application (PC only).

"There ought to be a law!" I hear you say. Funny you should mention it. *H.R. 5304: Preventing Harassment through Outbound Number Enforcement Act* was introduced back in 2006. It passed in the U.S. House of Representatives by a voice vote and was received in the Senate later that year. It never became law.

S. 2630: Truth in Caller ID Act of 2006—"a bill to amend the Communications Act of 1934 to prohibit manipulation of caller identification information"—was also introduced that year. It never became law either.

S. 30: Truth in Caller ID Act of 2009—"a bill to amend the Communications Act of 1934 to prohibit manipulation of caller identification information"—passed in the Senate "by unanimous consent" on February 23, 2010. It is now awaiting a House vote. Coincidentally, *H.R. 1110: PHONE Act of 2009*—"to amend title 18, United States Code, to prevent caller ID spoofing, and for other purposes"—passed in the House by roll-call vote on December 16, 2009, and is waiting for a Senate vote. These last two bills will also die if they are not passed before 2011.

Do you detect a pattern? Most efforts to pass protective laws have been met with inaction. Meanwhile, people are being spoofed and harassed, and the spoof-meisters are making money.

Ever wonder how businesses such as banks deal with the spoofing problem? They have to use counterspoofing services like TrustID.com. As that company explains it:

> The TrustID Telephone Firewall™ software as a service solution classifies calls between a calling party and a contact center as *authenticated*, *unauthenticated* or *high-probability fraudulent*, using real-time telephone network forensics. By validating the callers Caller ID and ANI's authenticity *before calls are answered*. The telephone firewall helps secure remote voice channels and make telephone-based commerce platforms more cost-effective.

"Do I have to pay to block caller ID?"

Myth. It costs money to block or unblock your caller ID when making a call.

Reality. The FCC protects your privacy rights. Phone

companies are required to provide a simple and uniform method of blocking and unblocking your Caller ID information on a call-by-call basis, at no charge. It is the feature that allows you to *receive* Caller ID information on your phone that may be leased for a fee. In some states, a charge for per-line blocking is allowed.

"When I block caller ID, does it keep my number private?"

Myth. Caller ID blocking keeps your phone number absolutely private.

Reality. Block all you want, but your number may still be transmitted when you call a toll-free number. (Hey, they are paying for the call, not you!) Also, your number will absolutely be transmitted when you dial 911, poison control services, or the emergency line of any public agency.

CALLER ID FACTS

The unwanted phone call as a phenomenon became a problem as soon as there were more than two phones in existence. Over the years, phone companies (pushed by customer and law enforcement demands) developed several solutions, such as operator intercept, "trap and trace," and caller ID unblocking. Although the solutions are effective, they require cooperation between the phone company, law enforcement, and a consumer who agrees to assist with a criminal prosecution.

Caller ID and call blocking are features the phone company developed for its customers who did not have access to the phone company's other number identification service, known as ANI (automatic number identification). ANI is the "caller ID" associated with toll-free phone numbers. It cannot be blocked or spoofed, as it is integral to the phone company billing system.

The somewhat crippled, untrustworthy caller ID system can still be useful. In addition to knowing how it works, however, you need to be aware of its limitations.

- Dialing *67 before you dial the number you want to reach blocks your phone number from being sent. It works only for that call. Every call you wish to block must be preceded by *67.
- If you have per-line blocking (where caller ID transmission is blocked by default), you may allow your phone number to be sent by dialing *82 before you dial a number you want to reach. Again, it works only for that call. Every call you wish to unblock must be preceded by *82.
- Some phone systems may offer additional privacy features, such as *61 caller ID block on, *65 caller ID block off, and *78 do not disturb.

Having caller ID problems? The FCC (http://www.fcc.gov/cib/consumerfacts/callerid.html) says:

> If you have caller ID and [you] receive a call from a telemarketer without the required caller ID information, if you suspect that caller ID information has been falsified, or you think the rules for protecting the privacy of your telephone number have been violated, you can file a complaint with the FCC. There is no charge for filing a complaint. You can file your complaint using an online complaint form found at esupport.fcc.gov/complaints.htm.

UNWANTED CALLS, TEXT MESSAGES & E-MAILS

Sometimes the problem is not spying via phone but rather harassment via phone. Sales calls, old flames' e-mails, and even weird text messages (many of which are designed to cost you money) are all problems seeking solutions. Thankfully, there are plenty of solutions available to you.

THE BASICS

We do not know if Albert Einstein ever said, "A bad haircut can make anybody look stupid." But, we do know this: Good security requires a good foundation. The following telephone security measures are basic. If you implement them first, you are on your way to giving your phone the power of the Death Star's energy shield, protecting it from rebel phone calls, e-mails, and text invasions. Some of these basics obviously apply to either your cell or landline phone. Some apply to both.

Contact the National Do Not Call Registry

Get your phone number added to the list by contacting 1-888-382-1222 or https://www.donotcall.gov/. If using the toll-free telephone number, you must call from the phone number you want to have listed. Registering your phone number here helps to screen out the direct marketers. The registry is run by the U.S. government's Federal Trade Commission. There is no charge to be listed, and your listing never expires. The registry is open to all home and cell phone numbers.

Contact your local phone company

Call the business office of your phone service provider, and add password protection to your phone account. Doing this will help prevent others from accessing your account via impersonation. Without this added layer of security, it is fairly easy for others to gain access to your records, change your account information, and add spy-friendly features to your phone.

Ask to have your account "flagged" with a note saying the company should immediately notify you if any unauthorized attempt is made to access your account.

Have the phone company unlist your phone number.

Take a moment to review your account with the representative to whom you are speaking. Double-check each charge and feature, one by one.

Make sure a feature like **remote call forwarding** is not active. This could allow someone to route your calls to a twin of your voice mail, where the calls could be screened *before* being forwarded to your real voice mail.

Make sure there are no extras, such as **off-premises**

extensions, simultaneous ringing to other phone numbers, or **extra phone lines** being listed at your address. These, along with **remote call forwarding,** are favorite surveillance tricks employed by estranged significant others and ex-business partners. If you use an alternate telephony carrier like Vonage or your local cable company for phone service, you need to probe further. Specifically ask what *additional* features their systems have that could impact your privacy.

Request information about your particular carrier's options for blocking unwanted calls. Some carriers offer a service that blocks incoming calls arriving without caller ID information. Your phone company may also let you create a **blacklist** of phone numbers from which you *never* want to receive calls. You may even be able to create times of the day when all calls are automatically shunted to voice mail.

Privacy features like **caller ID blocking** and **anonymous call rejection** are charged differently by each carrier. Some are free; some are not. It's best to check with your particular carrier on this.

FEE-BASED SERVICES

Many fee-based phone privacy services have risen to meet consumer demand for telecommunications privacy. Here are just a few of them and what they can do to improve your cell phone privacy.

Blocking unwanted phone calls

Tossabledigits.com allows you to *never* give your phone number out to *anybody* (except Tossabledigits.com). For example, say you purchase a prepaid phone and you pay with cash. In theory, *you* are the only person who knows your phone number at this point. Even the phone company does not know you by name. All they know is that the number is being used by a particular cell phone, somewhere. Next, instead of giving your new phone number to others, you contact Tossabledigits.com and purchase one or more virtual phone numbers. These are the numbers you selectively give out. All calls placed to these virtual numbers will be covertly routed to your real phone. Here comes the bonus: You can custom control how you handle calls from each of the numbers, using features such as these:

- **Do not disturb.** Sends the call to voice mail.
- **Advanced voice mail.** Forwards messages to you via e-mail, text, or MP3 voice files. You can also call in or use the website to retrieve messages.
- **Call screening.** Allows you to hear who is calling by asking the person to identify himself by recording his/her name. If you decline the call, the person gets a free visit to your voice mail jail, without even knowing he/she has been rejected.

The service may also be used for distributing phone numbers with a limited life span. You may need to talk to a life insurance salesperson about a specific issue, but you don't want your phone number to be sold to the telemarketing zombies. Not so sure about that new acquaintance you met at your friend's party? Give the person tossable digits when he/she asks for your number. One false move, and this new acquaintance is dialing into thin air. Need a one-time number for your Craigslist or eBay ad? Use virtual digits.

Trapcall.com offers many privacy services. The company's pitch says, "TrapCall unmasks blocked and restricted calls, blacklists harassing callers, and can even record your incoming calls. There is no software to install, and it works on any mobile phone." As helpful as this service may appear to be, you need to know its current limitations. It works only with cell phones, not home landline phones, and it does not work with some cell phone carriers. Also, using the service will double your minutes usage because of the way it accomplishes its tasks.

Called.in offers phone number verification and reverse phone verification service via a Web-based process. Called.in acts as a middleman who passes a code number between you and the phone number you want to verify. While probably more useful to businesses than to individuals, it is still worth knowing that the service exists.

Blocking text/SMS messages

Blocking texts is a separate issue from blocking phone calls. To complicate matters further, not all unwanted text messages are alike. Some may be sent to you by people you know, from their phones. Other unwanted text messages come from spammers via the Internet. All text messages cost you money. As one carrier says, "[This phone company] bills for all messages whether sent or received, read or unread, solicited or unsolicited."

Unwanted texts from people you know are the easiest to handle. Directions for how to block these text messages can be found on the Internet by using the search phrase "block text messages" + "[name of cell phone carrier]." All major carriers have succinct instructions, which basically read like this:

To block unwanted text messages:

- Log in to Text Messaging with your registered wireless number.
- Once logged in, select Preferences; then choose Text Blocking from the menu on the left.
- Select the type of messages you want to block, or enter the specific addresses you want to block; then click Save.

Many popular cell phone operating systems (iPhone, BlackBerry, Android) also have some blocking features built into them.

Most spam messages are sent to your phone via your carrier's text address suffix (@txt.att.net, @vtext.com, etc.) with your ten-digit phone number as the prefix. The spammers are not targeting you specifically. They just use programs to generate random phone numbers and attach them to your carrier's suffix. Most will not be deliverable. However, some of the addresses these programs create will be for active phone numbers, possibly yours. If the message does not bounce, it is assumed that the address is valid and is then added to a "hot list." Next thing you know, you are being targeted by the spammers.

With many carriers, you have the choice of whether or not to receive messages via their Internet address. The problem is, if you turn off the Internet feature, you also block friendly messages from reaching you. No friends? No problem. But since most of us want to use our phone's communications features, we need a better solution.

TIP ▪ Take a page out of the Spy School 101 textbook, and ask your phone carrier for an alias address.

An **alias address** is a custom word, phrase, or number that replaces your ten-digit telephone number. Instead of your normal 9085551515@txt.att.net address, you can arrange it so you only receive messages sent to your alias address, nospam2me@txt.att.net. Then, give your alias address to your "in" crowd and block everything coming to your old telephone number address.

Each carrier handles this a bit differently. Your first step is to log in to your account and look for the applicable feature settings and directions for use.

Keep in mind that you may have more than one carrier address to block. For example, AT&T uses two suffixes: "txt.att.net" and "mms.att.com." Determine which of your carrier's addresses are applicable to your phone. In most cases, smart cell phones will have one address for text messages and another for MMS messages.

Blocking unwanted e-mail

Protecting yourself from unwanted e-mail such as spam has become routine for all of us, and these basic recommendations apply.

- Stop using your private e-mail address for general Internet use.
- Set up a spam filter on your mail server.
- Consider using a special e-mail address to receive messages on your cell phone.

There are many more spam-fighting recommendations and tricks available to help you. One of the better recommendations lists appears at http://www.wikihow.com/Stop-Spam.

CONCLUSION

To most of us, it is amazing that a manufacturer would brag about sales of its cell phone spyware. But what is the reality? Is the market for cell phone spyware really *that* big? One would think that *everyone* would be against it. After all, who doesn't value their privacy?

To find out, let's look at two recent surveys.

SURVEY NO. 1

An unscientific Internet poll was run for several months on spybusters.blogspot.com, a site whose visitors are very interested in things like privacy and electronic surveillance. Conservative results were expected.

The question: "Want a law against cell phone spyware?"

About 25 percent of people polled said, "No," and another 10 percent hedged and said, "Only for phones that one does not own." The poll results backed up spyware manufacturers' claims that people want this type of product! More than one-third of the poll population indicated they wanted free access to cell phone spyware for phones they own. And 25 percent of them want the ability to place spyware on their and everyone else's phones.

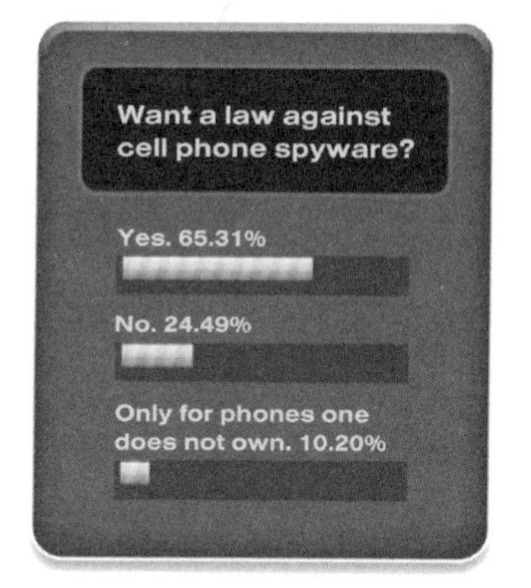

SURVEY NO. 2

Monster Worldwide, Inc. (an Internet employment site), recently asked its U.S. visitors some questions to see how they felt about their bosses.[14] One of the questions posed was this: If they could spy or eavesdrop on their boss without getting caught, would they?

Of the 2,153 respondents, 57 percent said they want to know what their bosses are saying about them behind closed doors. Only 12 percent said they would *not* eavesdrop on their boss because they are afraid of what they might hear.

As these survey results show, the question is not, "Who

14 Monster Worldwide, "'Nightmare' Bosses Not a Concern for Most Workers, According to New Monster.com Poll," http://www.about-monster.com/content/nightmare-bosses-not-concern-most-workers-according-new-monstercom-poll, September 16, 2010.

doesn't value their privacy?" The results scream, "Who doesn't value *your* privacy?"

Responses like the ones tallied in these two surveys show that we are not paranoid. Between news articles, lawsuits, manufacturers' sales claims, websites, and Internet polls, the picture becomes clear. The potential cell phone spyware market *is* huge. It is also incredibly profitable for software sellers and, indirectly, for the phone companies. This evidence leads to one conclusion: We as individuals will have to be responsible for protecting our own privacy.

As mentioned at the outset of this book, privacy invasion via technology that we do not fully understand is scary. Understanding, however, is simply a matter of education. You have wisely chosen education instead of cowering in ignorance.

The tips you've learned here make you stronger than your adversary. You know the truth about your communications technology. You know the truth about related privacy vulnerabilities. You know the truth about how to protect yourself. Your decision to become educated has given you this knowledge, and you have proven the old maxim: *Knowledge is power.* Congratulations, and best wishes for a happier life.

THE FUTURE

You can be very sure that smart cell phones are coming. The term *electronic wallet* is discussed freely, and rightfully so. We already use our *smart* phones for boarding passes at airports, and as tickets for live events. Future uses will include the ability to complete almost all monetary transactions.

With the addition of biosensors and advanced recognition apps, your phone is also destined to become a replacement for your keychain and your Certified Legal Identity Card (CLIC).

Manufacturers have more incentive than ever before to improve product security. The opportunities for rushing an insecure product to market for monetary gain are waning. That balance is changing. Futuristic features won't work without security. This does not mean, however, that the consumer will need to be any less cautious. Security is impotent unless activated. (Have you password protected your phone yet?)

The term "cell phone" now sounds as quaint as "dialing the phone." Given the mass consolidation of functionality, *Multi Electronic Memory & Expediter (MEME)* might be a better moniker for our future devices. ME-ME, it has a warm personal "ring" to it.

ADDITIONAL ASSISTANCE

Murray Associates provides advanced eavesdropping detection audits and counterespionage consulting services to businesses, government organizations, and high-profile individuals. Headquartered near New York City, Murray Associates makes its services available internationally.

Inquire online via www.spybusters.com/inquiry.html or by mail to Murray Associates, P.O. Box 668, Oldwick, NJ 08858 (USA). There is also a good chance our FAQ page at www.spybusters.com/TSCM_FAQ.html will have the answers you need.

Cases are accepted based on our schedule and our ability to help in each particular situation. Cases involving detection of legal eavesdropping or assistance with illegal activities, or which are against the best interests of the U.S. government and its citizens, are not accepted.

More personal counterespionage tips can be found in our other tutorials . . .

- www)) Quit Tapping Me: How You Can Find Residential Wiretaps Yourself
- www)) Business Spies . . . and the Top 10 Spy-Busting Tips They Don't Want You to Know
- www)) Countering Electronic Espionage in Business (FREE)
- www)) The Murray Manual (information about our services—FREE)

. . . and at www.spybusters.com/Cell911.html.

ABOUT THE AUTHOR

Kevin D. Murray is an independent, professional security consultant. He has been solving electronic eavesdropping, security, and counterespionage matters since 1973 while with Pinkerton's Inc., and from 1978 to present at his consulting firm, Murray Associates, which provides advanced eavesdropping detection (technical surveillance countermeasures, or TSCM) and counterespionage consulting services to business, government, and high-profile individuals.

Headquartered in the New York metropolitan area, with services available worldwide, Murray Associates invites inquiries from corporate, government, and professional security entities. Murray Associates' client family includes:

- 375+ Fortune 1000 companies
- 800+ others from every imaginable corner of business
- Many North American government agencies
- Clients in 39 states and several foreign countries

Mr. Murray's professional certifications are as follows:

- CISM (Certified Information Security Manager)
- CPP (Certified Protection Professional)
- CFE (Certified Fraud Examiner)
- BCFE (Board Certified Forensic Examiner)
- MPSC (Mobile Phone Seizure Certification)

His professional affiliations are listed below:

- International Association of Professional Security Consultants (IAPSC)
- Information Systems Audit & Control Association (ISACA/CISM)
- American Society for Industrial Security (ASIS/CPP)
- The American College of Forensic Examiners (ACFE/BCFE)
- High Technology Crime Investigation Association (HTCIA)
- Association of Certified Fraud Examiners (ACFE/CFE)
- Espionage Research Institute (ERI—Advisor)
- Infraguard (An FBI–Private Sector Initiative)

Mr. Murray also has the following licenses:

- Private Detective (New Jersey)
- Electronic Countermeasures/Counterintelligence (North Carolina)
- Amateur Radio Operator (FCC)
- LTA Pilot/Instructor (FAA)

Mr. Murray has developed and taught two Electronic Eavesdropping Detection seminars at John Jay College of Criminal Justice in New York City. He has provided technical advice to HBO, George Clooney, Steven Soderbergh, the Discovery Times Channel, Discovery Channel/Canada, ABC News' *20/20,* FOX News, CNN, CBS News, Joe Finder, NBC's *Dateline,* James Cameron, Orion Pictures . . . and others.

Mr. Murray is the author of several textbook chapters, white papers, and magazine articles (see the list at the end of this section).

His professional work has also been featured in books and magazines, as well as on radio and television. He is referenced and quoted by *Fortune* magazine, the *New York Times, USA Today,* NPR, International Security News, Corporate Security, Security Management, *Congressional Quarterly,* Security Letter, *Time,* American Public Radio, Boardroom Reports, the *Washington Post,* Business Finance, and several Internet security websites. Murray Associates has appeared on an episode of the HBO original series *K Street* with James Carville; the Discovery Times Channel documentary *Someone's Watching*; FOX News' *212,* (episode: "A Day with the Spybusters"); and "Spies Like Us," an episode of Public Radio International's (PRI) *This American Life* series with Ira Glass.

In his spare time Mr. Murray enjoys traveling, and he is currently learning how to crack an 8-foot kangaroo-leather bullwhip. Eventually, he hopes to be able to play his theremin without scaring the neighbors.

Visit Murray Associates at www.spybusters.com and keep up to date with the latest eavesdropping and espionage news at **Kevin's Security Scrapbook** (spybusters.blogspot.com). The companion Web page for this book is located at www.spybusters.com/Cell911.html.

BOOKS FEATURING KEVIN D. MURRAY:

Arrington, Winston. *Now Hear This! Electronic Eavesdropping Equipment Designs.* Chicago, IL: Sheffield Electronics Co., 1997. Electronic Countermeasures sections. (ISBN B0012K8WKW)

Bottom, N. R., and R. R. J. Gallati. *Industrial Espionage: Intelligence Techniques & Countermeasures.* Boston: Butterworth-Heinemann Ltd., 1984. Electronic Counterintelligence section. (ISBN 0409951080)

Guindon, Kathleen M. *A.M. Best's Safety & Security Directory.* Oldwick, NJ: A.M. Best & Co., 2001. Chapter 15, "Spy vs. Spy: Everything You Need To Know About Corporate Counterespionage." (LoC Catalog Card Number 74-618599, ISBN B000VU9PI2)

Johnson, William M. *101 Questions & Answers About Business Espionage.* N.p., Shoreline, Washington: The Questor Group, 2003. Questions and Answers section. (ISBN 1591096227)

Krieger, Gary R. *Accident Prevention Manual: Security Management.* N.p., Itasca, Illinois: National Safety Council, 1997. Chapter 20, "Communications Security." (ISBN 087912198X)

Lee, Edward L. II. *Staying Safe Abroad: Traveling, Working & Living in a Post-9/11 World.* Williamsburg, MI: Sleeping Bear Risk Solutions LLC, 2008. Eavesdropping Detection section. (ISBN 0981560504)

Mars-Proietti, Laura. *The Grey House Safety & Security Directory*. Amenia, NY: Grey House Publishing, 2004. Chapter 16, "Eavesdropping Detection." (ISBN 1592370675)

Montgomery, Reginald J., and William J. Majeski, eds. *Corporate Investigations*. Tucson, AZ: Lawyers & Judges Publishing, 2001 & 2005. Chapter 5, Electronic Eavesdropping & Corporate Counterespionage. (ISBN 0913875635)

Murray, Kevin D. Business Spies . . . and the Top 10 Spybusting Tips They Don't Want You to Know! Oldwick, NJ: Spybusters, LLC, 2010. (ISBN B003VYCDRK)

Murray, Kevin D. Electronic Eavesdropping & Industrial Espionage: The Missing Business School Courses. Oldwick, NJ: Spybusters, LLC, 1992–2010.

Reid, Robert N. *Facility Manager's Guide To Security: Protecting Your Assets*. Lilburn, GA: Fairmont Press, 2005. Chapter 12. (ISBN 0881734799)

Rothfeder, Jeffrey. *Privacy for Sale: How Computerization Has Made Everyone's Private Life an Open Secret*. New York: Simon & Schuster, 1992. Chapter 9, "Shadow of Technology." (ISBN 067173492X)

Schnabolk, Charles. *Physical Security: Practices & Technology*. Boston: Butterworth-Heinemann, 1983. Eavesdropping & Countermeasures chapter. (ISBN 040995067X)

Sennewald, Charles A. CPP. *Security Consulting*. Boston: Butterworth-Heinemann, 2004. Chapter 15, "A Successful Security Consulting Business Stands on a Tripod." (ISBN 0750677945)

Shannon, M. L. *The Phone Book: The Latest High-Tech Techniques and Equipment for Preventing Electronic Eavesdropping, Recording Phone Calls, Ending Harassing Calls, and Stopping Toll Fraud.* Boulder, CO: Paladin Press, 1998. Sidebars and illustrations. (ISBN 0873649729)

Shannon, M. L. *The Bug Book: Wireless Microphones & Surveillance Transmitters.* Boulder, CO: Paladin Press, 2000. (ISBN 1581600658)

Swift, Theodore N. *Wiretap Detection Techniques: A Guide To Checking Telephone Lines For Wiretaps.* Austin, TX: Thomas Investigative Publications, Inc., 2005. (ISBN 0918487056)

Walsh, Timothy J., and Richard J. Healy. *The Protection of Assets Manual.*[15] Aberdeen, WA: Silver Lake, 1987. Section 15, "Electronic Eavesdropping Detection." (ISBN 0930868048)

PERIODICALS FEATURING KEVIN MURRAY:

- *The Legal Investigator*
- *PI Magazine* (cover story)
- *Security Management* (ASIS)
- *World Association of Detectives News*

. . . and more.

15 This multivolume reference is often referred to as the "bible of the security industry."

SPYWARN MOBILE™ COUPON

A FREE SpyWarn Mobile™ transmission detector is available to everyone who purchases this book new (printed or as an e-book). If your bookseller was not able to include it with the book, you may request to receive it by mail. Simply fill out the coupon on the next page and send it in. Only original coupons can be honored, for obvious reasons.

If you purchased the e-book version, you may order your SpyWarn Mobile™ using our online form at http://www.spybusters.com/spywarn.html. Simply enter your address information, the seller's name, and the invoice number from your purchase receipt. Please enter the invoice number carefully. It must match our sales records from the seller.

If this is a used book and the coupon is missing, you may still obtain one or more SpyWarn Mobile™ device by using the PayPal banner at http://www.spybusters.com/spywarn.html.

SpyWarn Mobile™ is guaranteed for three months—ample time to help determine if your phone is bugged. A free replacement is offered during this time period should any non-battery failure occur. Simply send us your old detector, along with your name and address, to P.O. Box 668, Oldwick, NJ 08858 (USA), and we will promptly send you a replacement.